T0310822

Monographs in

Volume 11

Analytic Methods in Number Theory

When Complex Numbers Count

Monographs in Number Theory ISSN 1793-8341

Published

Vol. 11 *Analytic Methods in Number Theory: When Complex Numbers Count*
by Wadim Zudilin

Vol. 10 *Recent Progress on Topics of Ramanujan Sums and Cotangent Sums Associated with the Riemann Hypothesis*
by Helmut Maier, Michael Th Rassias & Lázló Tóth

Vol. 9 *Modular and Automorphic Forms & Beyond*
by Hossein Movasati

Vol. 8 *Topics and Methods in q-Series*
by James Mc Laughlin

Vol. 7 *The Theory of Multiple Zeta Values with Applications in Combinatorics*
by Minking Eie

Vol. 6 *Development of Elliptic Functions According to Ramanujan*
by K. Venkatachaliengar, edited and revised by Shaun Cooper

Vol. 5 *Hecke's Theory of Modular Forms and Dirichlet Series (2nd Printing with Revisions)*
by Bruce C. Berndt & Marvin I. Knopp

Vol. 4 *Number Theory: An Elementary Introduction Through Diophantine Problems*
by Daniel Duverney

Vol. 3 *Analytic Number Theory for Undergraduates*
by Heng Huat Chan

Vol. 2 *Topics in Number Theory*
by Minking Eie

Vol. 1 *Analytic Number Theory: An Introductory Course*
by Paul T. Bateman & Harold G. Diamond

Monographs in Number Theory
Volume 11

Analytic Methods in Number Theory

When Complex Numbers Count

Wadim Zudilin
Radboud University Nijmegen, The Netherlands

World Scientific
NEW JERSEY • LONDON • SINGAPORE • BEIJING • SHANGHAI • HONG KONG • TAIPEI • CHENNAI

Published by

World Scientific Publishing Co. Pte. Ltd.
5 Toh Tuck Link, Singapore 596224
USA office: 27 Warren Street, Suite 401-402, Hackensack, NJ 07601
UK office: 57 Shelton Street, Covent Garden, London WC2H 9HE

Library of Congress Control Number: 2023037907

British Library Cataloguing-in-Publication Data
A catalogue record for this book is available from the British Library.

Monographs in Number Theory — Vol. 11
ANALYTIC METHODS IN NUMBER THEORY
When Complex Numbers Count

ISBN 978-981-127-931-7 (hardcover)
ISBN 978-981-127-932-4 (ebook for institutions)
ISBN 978-981-127-933-1 (ebook for individuals)

For any available supplementary material, please visit
https://www.worldscienti ic.com/worldscibooks/10.1142/13498#t=suppl

Desk Editor: Rok Ting Tan

Typeset by Stallion Press
Email: enquiries@stallionpress.com

Printed in Singapore

To my family,
friends and teachers

Preface

There is no surprise that arithmetic properties of integral ('whole') numbers are controlled by analytic functions of complex variable. At the same time, the values of analytic functions themselves happen to be interesting numbers, for which we often seek explicit expressions in terms of other 'better known' numbers or try to prove that no such exist. This natural symbiosis of number theory and analysis is centuries old but keeps enjoying new results, ideas and methods. The present book takes a semi-systematic review of analytic achievements in number theory ranging from classical themes about primes, continued fractions, transcendence of π and resolution of Hilbert's seventh problem to some recent developments on the irrationality of the values of Riemann's zeta function, sizes of non-cyclotomic algebraic integers and applications of hypergeometric functions to integer congruences. Our principal goal is to present a variety of different analytic techniques that are used in number theory, at a reasonably accessible — almost popular — level, so that the materials from this book can suit for teaching a graduate course on the topic or for a self-study. Exercises included are of varying difficulty and of varying distribution within the book (some chapters get more than other); they not only help the reader to consolidate their understanding of the material but also suggest directions for further study and investigation. Furthermore, the end of each chapter features brief notes about relevant developments of the themes discussed.

Rome was not built in a day. One needs to get comfortable about the concept of using analytic tools in number theory, and this serves as a good reason for going first through the topics that are traditionally represented in books on analytic number theory. This is the principal task of Chapters 2, 4 and 5, for which complementing (or alternative) sources are existing books [4, 9, 21, 33, 36, 47, 59]. Chapters 2, 5 and 6 are particularly close to the exposition in [33] (never translated into English); this was my first textbook on the subject and a great influence on my personal perception. In addition, an inspiration for Chapter 8 came from a series of lectures by Yu. Nesterenko on transcendental numbers held in the 1990s, while Chapter 4 is a tribute to the late A. van der Poorten and his revolutionary simplistic treatment of continued fractions (the book [14] is a recommended complement to the chapter). The tone of Chapter 3 is more analytic, parts are rooted from the classical Whittaker–Watson textbook [80]. The choice of topics in the remaining chapters is guided by my personal tastes.

My close family (*het gezin* in Dutch) has been a lasting inspiration for my academic career, of which this book is a tiny outcome. Thank you, Olga and Victor, for your constant support and encouragement during the years!

It is my pleasure to thank friends and colleagues whose valuable feedback helped to improve the text: Heng Huat Chan, Björn Johannesson, Pieter Moree, Berend Ringeling, and Armin Straub. The staff and editors at the World Scientific assisted me at all stages of transmission of my manuscript into the publication, and I am particularly thankful to Rok Ting Tan for making this book possible.

The last but not least thing is to thank the reader who follows the book in either physical or electronic format. Enjoy!

Wadim Zudilin
Nijmegen (NL) and Newcastle (AU)
April 2023

Contents

Chapter 1

Numbers and q-numbers

1.1 Prime numbers

The task of finding the greatest common divisor (a, b) of two integers a and b is traditionally performed using the Euclidean algorithm (or its numerous extensions). The following statement is a usual companion of the algorithm, which we review (in a somewhat friendlier context) in Chapter 4. Our proof below is less efficient but works well and is simpler from a theoretical point of view.

Lemma 1.1. *Suppose that two positive integers a and b are relative prime, $(a, b) = 1$. Then the (diophantine) equation $ax + by = 1$ is solvable in integers x, y.*

Proof. Perform mathematical induction on $a + b$. The base of induction corresponds to $a + b = 2$, hence $a = b = 1$, in which case we can take $x = 1$ and $y = 0$ as the solution of $ax + by = 1$. Assume that the statement is true for $a + b < r$ and consider the situation $a + b = r$. Since $(a, b) = 1$, assuming without loss of generality that $a > b$ we also have $(a - b, b) = 1$ so that $(a - b)u + bv = 1$ has a solution $u, v \in \mathbb{Z}$ because $(a - b) + b = a < r$. But then the pair $x = u$, $y = v - u$ solves $ax + by = 1$. $\square$

Recall that a positive integer $p \geq 2$ is said to be *prime* if all its positive integer divisors are exhausted by 1 and p itself.

Lemma 1.2. *If a product of several multiples is divisible by a prime p then at least one of the multiples is divisible by p.*

Proof. Without loss of generality consider the case of a product of just two multiples a and b. Given ab is divisible by p, we either have $p \mid a$ (and then the required statement follows) or p is coprime to a. In the latter case, $(a,p)=1$ so that $ax+py=1$ for some $x,y\in\mathbb{Z}$ by Lemma 1.1. Multiplying the both sides of the equality by b we obtain $abx+pby=b$. The prime p divides both the summands on the left-hand side because $p\mid ab$ (by the hypothesis) and $p\mid p$ for trivial reasons. It follows then that p divides the right-hand side, that is, b. □

Lemma 1.3. *The least divisor $a\ge 2$ of an integer $n\ge 2$ is a prime number.*

Proof. Assuming on the contrary that a is not prime, we conclude that is possesses a smaller divisor $b\ge 2$. Then $b\mid a$ and $a\mid n$ implies $b\mid n$, so that $b\ge 2$ is a smaller divisor of n than a. Contradiction that leads to the desired conclusion. □

Theorem 1.1 (fundamental theorem of arithmetic). *Any integer greater than 1 is decomposed into a product of primes, with possible repetitions, and this decomposition is unique up to permutation of the primes in the product.*

Proof. Let $n\ge 2$ be given. Take $p_1>1$ its least divisor; it is prime by Lemma 1.3, and so $n=p_1n_1$ for some $n_1\ge 1$. If $n_1=1$ then we already have the decomposition of n into a product of primes; otherwise, apply the procedure to $n_1\ge 2$ to get $n_1=p_2n_2$ etc. Since $n_1<n$, this process can be officially done through the mathematical induction.

The uniqueness is performed in a similar fashion, using Lemma 1.2 as the principal ingredient. Assume there is an $n>1$ for which two representations $n=p_1\cdots p_s$ and $n=q_1\cdots q_r$ exist, where all p_i and q_j are primes, and choose n the least positive integer with this property. It follows from the first representation that $p_1\mid n$ so that the product $q_1\cdots q_r$ is divisible by p_1, hence at least one of the multiples q_j is divisible by p_1. By rearranging the order we can consider q_1 divisible by p_1. But q_1 is prime, therefore $q_1=p_1$. Now we cancel the factor $p_1=q_1$ and denote $n_1=p_2\cdots p_s=q_2\cdots q_r$. Since $n_1<n$, the decomposition of n_1 into a product of primes is unique, so that $r=s$ and $p_2,\dots,p_s$ is a rearrangement of $q_2,\dots,q_s$. This implies that $n=p_1\cdots p_s=q_1\cdots q_r$ are same representations, which contradicts to our choice of n. □

A standard way to write the decomposition of a positive integer n into the primes is

$$n=p_1^{\alpha_1}p_2^{\alpha_2}\cdots p_s^{\alpha_s}, \tag{1.1}$$

where $p_1, \dots, p_s$ are pairwise distinct primes and $\alpha_1, \dots, \alpha_s$ are positive exponents.

Exercise 1.1. Show that all positive divisors of n are exhausted by the list

$$\{d = p_1^{\beta_1} \cdots p_s^{\beta_s} : 0 \le \beta_1 \le \alpha_1, \dots, 0 \le \beta_s \le \alpha_s\}$$

and compute the total number $\tau(n)$ of the divisors and the sum $\sigma(n)$ of the divisors.

The arithmetic functions $\tau(n)$ and $\sigma(n)$ from the exercise are examples of multiplicative functions we will meet in Section 2.4. Another example is the Möbius function $\mu(n)$ defined by $\mu(p_1^{\alpha_1} p_2^{\alpha_2} \cdots p_s^{\alpha_s}) = (-1)^s$ if $\alpha_1 = \alpha_2 = \cdots = \alpha_s = 1$ and by 0 otherwise (in other words, when n is *not square-free*, that is, divisible by a square of some integer $m > 1$); conventionally, $\mu(1) = 1$ as an empty product is always understood as 1.

Exercise 1.2. (a) Prove that

$$\sum_{d|n} \mu(d) = \begin{cases} 1 & \text{if } n = 1, \\ 0 & \text{otherwise.} \end{cases}$$

(b) Prove the *Möbius inversion formula*: if

$$F(n) = \sum_{d|n} f(d)$$

then

$$f(n) = \sum_{d|n} \mu(n/d) F(d) = \sum_{d|n} \mu(d) F(n/d).$$

In a slightly different arithmetic direction, we can use the explicit structure of divisors of natural numbers to control the prime decomposition of the factorial $n! = 1 \cdot 2 \cdot 3 \cdots n$.

Exercise 1.3. (a) Let

$$n! = \prod_{p \le n} p^{\nu_p}$$

be the canonical decomposition of $n!$ into the product of primes. Show that

$$\nu_p = \operatorname{ord}_p(n!) = \left\lfloor \frac{n}{p} \right\rfloor + \left\lfloor \frac{n}{p^2} \right\rfloor + \left\lfloor \frac{n}{p^3} \right\rfloor + \cdots.$$

(Here $\lfloor x \rfloor$ denotes the integer part of a real number x, that is, $\lfloor x \rfloor \le x < \lfloor x \rfloor + 1$, and the sum terminates since the terms for which $p^k > n$ contribute trivially.)

(b) Show that the very same ν_p can be computed via the formula $(n - S_p(n))/(p-1)$, where $S_p(n)$ denotes the sum of digits in n written in base p.

We return to the prime number theme in greater detail in Chapter 2 for investigating the asymptotic distribution of primes.

1.2 Integer-valued factorial ratios

There are many ways to define the binomial coefficients $\binom{n}{m}$, for example, as the quantities appearing in the binomial theorem

$$(a+b)^n = \sum_{m=0}^{n} \binom{n}{m} a^m b^{n-m} \tag{1.2}$$

or, more pragmatically, via the explicit formula

$$\binom{n}{m} = \frac{n!}{(n-m)!\,m!}, \quad \text{where } n \geq m \geq 0.$$

It is a fundamental fact that these ratios of factorials are integers. There are several ways of demonstrating this.

Analytical proof. Use the Pascal triangle relations

$$\binom{n}{m} = \binom{n-1}{m} + \binom{n-1}{m-1}$$

and mathematical induction.

Linear algebra proof. Use the generating function

$$\sum_{m=0}^{n} \binom{n}{m} t^m = (1+t)^n = \underbrace{(1+t)\dots(1+t)}_{n \text{ times}} \in \mathbb{Z}[t].$$

Combinatorial proof. $\binom{n}{m}$ counts the number of m-element subsets of an n-set.

Arithmetic proof. The order in which a prime p enters $n!$ is computed in Exercise 1.3. Setting $x = (n-m)/p^k$ and $y = m/p^k$ in the inequality

$$\lfloor x+y \rfloor - \lfloor x \rfloor - \lfloor y \rfloor \geq 0,$$

and summing k over the positive integers, we see that

$$\operatorname{ord}_p \binom{n}{m} = \operatorname{ord}_p(n!) - \operatorname{ord}_p(m!) - \operatorname{ord}_p((n-m)!) \geq 0 \quad \text{for any prime } p.$$

You may find the last strategy most sophisticated. But it is truly arithmetic!

Exercise 1.4. The Catalan numbers are officially defined by

$$C_n = \frac{1}{n+1}\binom{2n}{n} \quad \text{for } n = 0, 1, 2, \dots .$$

(a) Prove that

$$C_{n+1} = \sum_{i=0}^{n} C_i C_{n-i} \quad \text{for } n \ge 0.$$

(b) Show that the Catalan numbers are integral for any $n \ge 0$.

Hint. You may choose neither to notice that $C_n = \binom{2n}{n} - \binom{2n}{n+1}$ nor to use part (a). □

Binomial coefficients are a source of many other integral ratios of factorials, like

$$\frac{(6n)!}{n!\,(2n)!\,(3n)!} = \binom{3n}{n}\binom{6n}{3n}$$

and

$$\frac{(5n)!^2}{n!\,(2n)!\,(3n)!\,(4n)!} = \binom{5n}{n}\binom{5n}{2n}.$$

There are however many other cases not reducible to binomials, like

$$\frac{n!\,(30n)!}{(6n)!\,(10n)!\,(15n)!}. \tag{1.3}$$

In 1850 Chebyshev used the integrality of these ratios to give a sharp upper bound for the prime counting function.

Exercise 1.5. (a) Prove that for $n > 0$ the number (1.3) cannot be represented as a product of binomial coefficients.

(b) Show that Chebyshev's numbers (1.3) are integral for any $n \ge 0$.

Hint. (b) Use Exercise 1.3 and the inequality

$$f(x) = (\lfloor 30x \rfloor + \lfloor x \rfloor) - (\lfloor 6x \rfloor + \lfloor 10x \rfloor + \lfloor 15x \rfloor) \ge 0 \tag{1.4}$$

valid for all real x. The function $f(x)$ assumes only integral values because of the way it is defined; you need to demonstrate that $f(x)$ is either 0 or 1 for all real x. In order to establish this, make use of $y = \lfloor y \rfloor + \{y\}$ where $\{y\}$ is the fractional part of y (which is a 1-periodic function), so that $f(x) = (\{6x\} + \{10x\} + \{15x\}) - (\{30x\} + \{x\})$ is 1-periodic and the required property of x is to be shown for the range $0 \le x < 1$ only. □

Chebyshev's example falls into the category of more general factorial ratios

$$D_n(\boldsymbol{a},\boldsymbol{b}) = \frac{(a_1n)!\cdots(a_rn)!}{(b_1n)!\cdots(b_sn)!}\,, \tag{1.5}$$

where the integer vectors $\boldsymbol{a} = (a_1,\dots,a_r)$ and $\boldsymbol{b} = (b_1,\dots,b_s)$ satisfy the balancing condition

$$\sum_{i=1}^{r} a_i = \sum_{j=1}^{s} b_j$$

(which we will always assume in what follows) and the arithmetic condition

$$\sum_{i=1}^{r} \lfloor a_i x\rfloor - \sum_{j=1}^{s} \lfloor b_j x\rfloor \ge 0 \quad \text{for } x \in \mathbb{R}. \tag{1.6}$$

By the arithmetic proof above or by the arithmetic argument for part (b) of Exercise 1.5 we conclude that condition (1.6) is sufficient and necessary for the integrality of $D_n(\boldsymbol{a},\boldsymbol{b})$. (Yes, indeed, it is necessary as well: if (1.6) does not hold then, for many values of n, there are primes that show up in the denominator of the corresponding $D_n(\boldsymbol{a},\boldsymbol{b})$ with larger exponent than in the denominator, because the corresponding analogue of function (1.4) assumes negative values for some $0 \le x < 1$.) This result [52] is known in the literature as *Landau's criterion.*

The same argument, based on the inequality

$$\lfloor 2x\rfloor + \lfloor 2y\rfloor - \lfloor x\rfloor - \lfloor x+y\rfloor - \lfloor y\rfloor \ge 0,$$

shows that the so-called *super-Catalan numbers*, or the *Gessel numbers* (after Gessel [35] who introduced them in 1992 with a combinatorial motivation)

$$G_{m,n} = \frac{(2m)!\,(2n)!}{m!\,(m+n)!\,n!}$$

are integers as well for all $m,n \ge 0$. Notice that when $m = 1$ this is just two times the ordinary Catalan number C_n. But as for binomial coefficients we can also prove that $G_{m,n} \in \mathbb{Z}$ using other techniques, for example, the identity

$$G_{m,n} = \sum_{k=-\min\{m,n\}}^{\min\{m,n\}} (-1)^k \binom{2n}{n+k}\binom{2m}{m+k},$$

due to von Szily (1894). The latter may be challenging for you to verify but it is *algorithmically provable* [61] meaning that there is no need for performing tedious calculations, which are required in the arithmetic argument. One can also prove the integrality of $G_{m,n}$ by induction using the initial conditions $G_{n,n} = G_{n,0} = \binom{2n}{n}$ and the formula

$$G_{m,m+\ell} = \sum_{k=0}^{\lfloor \ell/2 \rfloor} 2^{\ell-2k} \binom{\ell}{2k} G_{m,k} \tag{1.7}$$

due to Gessel [35]. The formulae of von Szily and Gessel originate from analysis and combinatorics; their analogues for general integer-valued factorial ratios $D_n(\boldsymbol{a}, \boldsymbol{b})$ are not known. This makes the arithmetic argument quite exclusive.

1.3 The world of q-numbers

The material of this section may be considered as advanced to the reader whose familiarity with polynomials, their reducibility and their zeros (often called roots) in finite extensions of the field of $\mathbb{Q}$ is still in a developing process. We touch these aspects in greater detail in Section 6.1; but already at this point it is useful to single out a particular family of monic polynomials — cyclotomic polynomials

$$\Phi_m(x) = \prod_{\substack{j=1 \\ (j,m)=1}}^{m} (x - e^{2\pi i j/m}). \tag{1.8}$$

The product in (1.8) is taken over *primitive roots of unity* $U_m = \{e^{2\pi i j/m} : j = 1, \dots, m,\ (j,m) = 1\}$ of degree m, that is, over those $\alpha \in \mathbb{C}$ for which $\alpha^m = 1$ but $\alpha^n \neq 1$ when $0 < n < m$. There are natural bijections (automorphisms) of the set U_m onto itself given by $\alpha \mapsto \alpha^a$, where integers a satisfy $(a, m) = 1$; the inverse bijection is given by $\alpha \mapsto \alpha^b$ with b such that $ab \equiv 1 \pmod{m}$. It is not hard to see that the set G_m of these automorphisms has a structure of group, which is isomorphic to the (multiplicative) group $(\mathbb{Z}/m\mathbb{Z})^* = \{a \pmod{m} : (a, m) = 1\}$. The latter group will be featured in many further discussions below; in particular, its order $\varphi(m) = |(\mathbb{Z}/m\mathbb{Z})^*|$ known as Euler's totient function will be computed. The automorphisms from G_m (and they only!) permute the roots of unity U_m, hence leave the cyclotomic polynomial $\Phi_m(x)$ unchanged; they form the Galois group of the polynomial. This explicit description immediately implies that (1.8) is a polynomial with *integral* coefficients, *irreducible* over $\mathbb{Z}$ and even over $\mathbb{Q}$ (so that any sub-product in (1.8) is not in $\mathbb{Q}[x]$), of degree $\varphi(m)$.

There is a natural context, in which these irreducible cyclotomic polynomials are viewed as polynomial analogues of prime numbers, more accurately, as deformed primes. For historical reasons, a traditional name for variable in this deformation is q rather than x, so that the construction below is a q-deformation of natural numbers.

Define q-numbers as

$$[n] = [n]_q = \frac{1-q^n}{1-q} = 1 + q + q^2 + \cdots + q^{n-1}$$

for $n = 1, 2, \ldots$. Clearly, $[n]_q \to n$ in the limit as $q \to 1$. The q-numbers are polynomials of degree $n-1$. Unlike the polynomials $\Phi_n(q)$ they are in general reducible over $\mathbb{Q}$. But because $q^n - 1 = \prod_{j=1}^n (q - e^{2\pi ij/n})$, we essentially know all irreducible factors of $[n]_q$: they are cyclotomic.

Exercise 1.6. (a) Show that, for $n = 2, 3, \ldots,$

$$[n]_q = \prod_{\substack{m|n \\ m>1}} \Phi_m(q). \tag{1.9}$$

(b) Use the Möbius inversion formula to conclude from part (a) that for the same range of n,

$$\Phi_n(q) = \prod_{\substack{m|n \\ m>1}} [m]_q^{\mu(n/m)}.$$

Formula (1.9) tells us, in particular, that a q-number $[n]_q$ is an irreducible polynomial if and only if n is prime, in which case $[n]_q$ coincides with $\Phi_n(q)$. But it also suggests that the formula is a q-deformation of the decomposition (1.1). This is a bit harder to believe to, since the $q \to 1$ limit in (1.9) results in

$$n = \prod_{\substack{m|n \\ m>1}} \Phi_m(1)$$

and this (more regular looking!) product does not really resemble (1.1). Nevertheless the result truly duplicates the latter (if we accept to ignore numerous ones that appear as values $\Phi_m(1)$), because of the following evaluations.

Exercise 1.7. (a) Let p be prime. Verify that, for $\alpha \ge 2$, we have $\Phi_{p^\alpha}(q) = \Phi_{p^{\alpha-1}}(q^p)$. In particular, $\Phi_{p^\alpha}(1) = p$ for any $\alpha \ge 1$.

(b) Show that, more generally,

$$\Phi_{pm}(q) = \begin{cases} \Phi_m(q^p) & \text{if } (m,p) = p, \\ \Phi_m(q^p)/\Phi_m(q) & \text{if } (m,p) = 1. \end{cases}$$

(c) Using part (b) or otherwise prove that $\Phi_m(1) = 1$ if $m > 1$ is not of the form p^α.

The next natural step in this story is defining q-factorials

$$[n]! = [n]_q! = \prod_{k=1}^{n} [k]_q,$$

whose irreducible polynomial factors are exclusively $\Phi_\ell(q)$ with $\ell = 2, 3, 4, \dots$, and the q-binomial coefficients

$$\begin{bmatrix} n \\ m \end{bmatrix} = \begin{bmatrix} n \\ m \end{bmatrix}_q = \frac{[n]!}{[m]!\,[n-m]!}$$

also known by name of Gaussian polynomials. Using (1.9) and arguing as in Exercise 1.3 we conclude that

$$\operatorname{ord}_{\Phi_\ell(q)} [n]! = \left\lfloor \frac{n}{\ell} \right\rfloor \quad \text{for all } \ell = 2, 3, 4, \dots . \tag{1.10}$$

In particular, this implies that $\left[\begin{smallmatrix} n \\ m \end{smallmatrix}\right] \in \mathbb{Z}[q]$. (You may also verify the q-Pascal triangle relations

$$\begin{bmatrix} n \\ m \end{bmatrix} = q^m \begin{bmatrix} n-1 \\ m \end{bmatrix} + \begin{bmatrix} n-1 \\ m-1 \end{bmatrix}$$

which together with the boundary conditions $\left[\begin{smallmatrix} n \\ 0 \end{smallmatrix}\right] = \left[\begin{smallmatrix} n \\ n \end{smallmatrix}\right] = 1$ lead to the same conclusion. There is also a q-deformation of the binomial theorem, which we touch in Chapter 10.)

More generally, we conclude that the q-versions of the Chebyshev–Landau ratios

$$D_n(\boldsymbol{a}, \boldsymbol{b}; q) = \frac{[a_1 n]! \cdots [a_r n]!}{[b_1 n]! \cdots [b_s n]!},$$

subject to the conditions $\sum_{i=1}^{r} a_i = \sum_{j=1}^{s} b_j$ and (1.6), are all polynomials in $\mathbb{Z}[q]$. This latter property, which we may call q-integrality via the analogy of what we have seen in Section 1.2, does not require from us any special effort; all we use are the *same* inequalities (1.6), while the formula for $\operatorname{ord}_p n!$ is replaced with the (somewhat simpler) formula for $\operatorname{ord}_{\Phi_\ell(q)} [n]!$. There is an interesting counterpart here for the polynomials $D_n(\boldsymbol{a}, \boldsymbol{b}; q)$ which is not seen by the numbers $D_n(\boldsymbol{a}, \boldsymbol{b}) = D_n(\boldsymbol{a}, \boldsymbol{b}; 1)$. We *expect* [79] that all $D_n(\boldsymbol{a}, \boldsymbol{b}; q)$ are not just polynomials in $\mathbb{Z}[q]$ but that they also have *non-negative* coefficients. Note that the coefficients of their irreducible cyclotomic factors $\Phi_\ell(q)$ when ℓ is not of the form p^α have both positive and negative coefficients; this is an easy consequence of Exercise 1.7. Therefore, the expectation about non-negativity is not trivial; in fact, we do not know how to prove it in the above generality, though some particular instances can be shown by rather elementary means. For example, the non-negativity of q-binomial coefficients follows from the q-Pascal triangle relations. Similarly, a q-version of the identity (1.7) allows one to show the non-negativity of the q-Gessel numbers

$$G_{m,n}(q) = \frac{[2m]!\,[2n]!}{[m]!\,[m+n]!\,[n]!}$$

for all $m, n \geq 0$.

Exercise 1.8. In this exercise we look at very simple rational functions

$$W_q(m,n;k) = \frac{(1-q^{mn})(1-q^k)}{(1-q^m)(1-q^n)} \quad \text{with } (m,n) = 1.$$

(a) Show that $W_q(m,n;k)$ are polynomials in $\mathbb{Z}[q]$.
(b) Show that the coefficients of $W_q(m,n;k)$ are non-negative if $k \geq (m-1)(n-1)$. Can this happen for some $k < (m-1)(n-1)$?

Chapter notes

Though we have not yet come across the Riemann hypothesis (RH), it is worth mentioning that the Chebyshev–Landau factorial ratios (1.5) show up naturally in its potential resolution, thanks to the equivalent Nyman–Beurling formulation. In relation with this, Bober [13] lists all such integral factorial ratios subject to the condition $s \leq r+1$ (which implies $s = r+1$). Furthermore, the very same ratios show up in arithmetic study of so-called mirror maps attached to Calabi–Yau manifolds and in characterisation of globally bounded hypergeometric series; none of these advanced topics is discussed in this book and therefore we do not even define them properly (but check equation (10.15) in Chapter 10 for a definition of generalized hypergeometric series). However Bober's analysis makes crucial use of a simple criterion, due to F. Rodríguez Villegas (2005), saying that the integrality (1.5) when $s = r+1$ is equivalent to the *algebraicity* of the corresponding generating series

$$\sum_{n=0}^{\infty} D_n(\boldsymbol{a}, \boldsymbol{b}) z^n$$

(which is a hypergeometric series), that is, the latter satisfies a solution of polynomial equation with coefficients in $\mathbb{Z}[z]$. Writing such a polynomial down is not plausible for most of the examples because the degree is large;

it is equal to 483 840 in Chebyshev's example ($a_1 = 1$, $a_2 = 30$, $b_1 = 6$, $b_2 = 10$, $b_3 = 15$). The expectation at the end of Section 1.3 raises naturally the question whether there are methods to prove the integrality of the Chebyshev–Landau factorial ratios without the arithmetic argument.

In Chapter 10 we return to the q-deformation as a main theme, so that we discuss it in a more analytic context (while still aiming at applications in number theory!). Here we would only highlight a q-deformation of a non-integral — in fact a *transcendental* number (as we will see in Chapter 6), namely the number π. Among many analytic expressions that define the quantity (and we witness some more in Chapter 10) we single out the classical representations

$$\pi = 4\sum_{n=0}^{\infty} \frac{(-1)^n}{2n+1} \tag{1.11}$$

due to Leibniz and

$$\pi = \left(\int_{-\infty}^{\infty} e^{-x^2}\,\mathrm{d}x\right)^2,$$

the Gaussian probability density integral. If we now define

$$\pi_q = 1 + 4\sum_{n=0}^{\infty} \frac{(-1)^n q^{2n+1}}{1-q^{2n+1}}, \quad |q| < 1,$$

then it is not hard to check the 'agreement' with Leibniz formula:

$$\lim_{\substack{q\to1\\|q|<1}} (1-q)\pi_q = \pi$$

(recall that $(1-q)/(1-q^{2n+1}) = 1/[2n+1]_q \to 1/(2n+1)$ and also $q^{2n+1} \to 1$ as $q \to 1$). It is more difficult (as requires some knowledge from the theory of modular forms) to convince yourself that for π_q so defined we have a different representation

$$\pi_q = \left(\sum_{n=-\infty}^{\infty} q^{n^2}\right)^2,$$

which is more in line with the Gaussian integral. Though such q-deformations are not unique when based on a single formula, those that fit several formulae are usually ones to consider; this strategy essentially dictates π_q to be essentially a canonical q-deformation of π.

Exercise 1.9. Show that

$$\pi_q = 1 + 4\sum_{m=1}^{\infty} \frac{q^m}{1+q^{2m}}.$$

Note that computing the limit of $(1-q)\pi_q$ as $q \to 1$ is challenging for this series representation.

Chapter 2

Prime number theorem

For the entire duration of this chapter we stick to the standard convention of using $\pi(x)$ for counting the prime numbers less than or equal to x:

$$\pi(x) = \sum_{p \le x} 1;$$

here $x > 0$ is a real number. For example, $\pi(1) = 0$, $\pi(10) = 4$, $\pi(10^{12}) = 37\,607\,912\,018$ and $\pi(p_n) = n$, where p_n denotes the nth prime. No exact formula for the function $\pi(x)$ is known, though resolution of the famous Riemann Hypothesis will give a reasonably simple way to legally compute it. Here is some historical information about development of our knowledge about the distribution of primes:

- Euclid: $\pi(x) \to \infty$ as $x \to \infty$;
- Euler: $\pi(x)/x \to 0$ as $x \to \infty$;
- Chebyshev (1848): if the limit

  $$\lim_{x \to \infty} \frac{\pi(x)}{x/\ln x}$$

  exists then it is equal to 1;
- Hadamard and de la Vallée-Poussin (1896): the asymptotic distribution of the prime numbers among the positive integers is given by

  $$\pi(x) \sim \frac{x}{\ln x} \quad \text{as } x \to \infty.$$

2.1 Chebyshev's bounds for primes

Lemma 2.1. *Let n be a positive prime and $K = \operatorname{lcm}(1, 2, \ldots, 2n+1)$ (the least common multiple). Then $K > 4^n$.*

Proof. Consider the integral

$$I = \int_0^1 x^n(1-x)^n \mathrm{d}x.$$

Since $0 < x(1-x) \le \frac{1}{4}$ on the interval $0 < x < 1$, we have $0 < I < 1/4^n$. On the other hand,

$$x^n(1-x)^n = a_n x^n + a_{n+1}x^{n+1} + \cdots + a_{2n}x^{2n}$$

for some *integers* $a_n, a_{n+1}, \ldots, a_{2n}$, so that the integration gives us

$$I = \frac{a_n}{n+1} + \frac{a_{n+2}}{n+2} + \cdots + \frac{a_{2n}}{2n+1}.$$

This implies that $K \times I$ is a positive integer; in particular, $KI \ge 1$. The latter estimate together with the bound $I < 1/4^n$ implies the claim. □

Lemma 2.2. *The product $\prod_{p\le x} p$ over primes is bounded from above by 4^x for each $x \ge 2$.*

Proof. It is sufficient to prove the statement for integral x, as then $\prod_{p\le x} p = \prod_{p\le\lfloor x\rfloor} p < 4^{\lfloor x\rfloor} \le 4^x$.

Use the mathematical induction on integer $x \ge 2$. The statement is clearly true for $x = 2, 3$. Assume that it is true for all $x < n$, where $n \ge 4$, and show that it also holds for $x = n$. If n is even then $\prod_{p\le n} p = \prod_{p\le n-1} p < 4^{n-1} < 4^n$. Therefore, let us concentrate on the case of odd n, so that $n = 2m-1$ for some $m \ge 3$. Split our product into two parts,

$$\prod_{p\le n} p = \prod_{p\le 2m-1} p = \prod_{p\le m} p \times \prod_{m<p\le 2m-1} p < 4^m \binom{2m-1}{m},$$

since all the primes from the second product divide the factorials in the numerator of the binomial coefficient

$$\binom{2m-1}{m} = \frac{(2m-1)!}{m!\,(m-1)!}$$

but do not divide the factorials in the denominator, so that Theorem 1.1 implies $p \mid \binom{2m-1}{m}$ for all $m < p \le 2m-1$. Using the binomial theorem in the form

$$2 \times \binom{2m-1}{m} = \binom{2m-1}{m} + \binom{2m-1}{m-1} < \sum_{k=0}^{2m-1} \binom{2m-1}{k} = (1+1)^{2m-1},$$

we deduce $\binom{2m-1}{m} < 2^{2m-2} = 4^{m-1}$. Thus, $\prod_{p\le n} p < 4^m \times 4^{m-1} = 4^n$ in the case of odd n as well. □

Theorem 2.1. *There exist absolute positive real constants a and b such that for all $x \geq 2$ we have*

$$a\,\frac{x}{\ln x} < \pi(x) < b\,\frac{x}{\ln x}.$$

Proof. We will prove the theorem with the constants $a = \frac{1}{2}\ln 2$ and $b = 6\ln 2$ for $x \geq 6$.

Choose n such that $2n+1 \leq x < 2n+3$ and take $K = \operatorname{lcm}(1, 2, \ldots, 2n+1) = p_1^{\alpha_1}\cdots p_s^{\alpha_s}$, where $s = \pi(2n+1)$. First notice that $p_i^{\alpha_i} \leq 2n+1$ for all i, as each $p_i^{\alpha_i}$ must appear on the list $1, 2, \ldots, 2n+1$ of the first natural numbers by the definition of the least common multiple. Therefore, $K = p_1^{\alpha_1}\cdots p_s^{\alpha_s} \leq (2n+1)^s$. On the other hand, from the estimate derived in Lemma 2.1 we find out that $(2n+1)^s > 4^n$. Taking the logarithm gives

$$\pi(x) \geq \pi(2n+1) = s > \frac{2n}{\log_2(2n+1)} > \frac{x-3}{\log_2 x} \geq \frac{x/2}{\log_2 x} = a\,\frac{x}{\ln x}.$$

Now proceed with the estimate from above. We have

$$\begin{aligned}\pi(x) &= \sum_{p\leq x} 1 = \sum_{p\leq\sqrt{x}} 1 + \sum_{\sqrt{x}<p\leq x} 1\\ &< \pi(\sqrt{x}) + \sum_{\sqrt{x}<p\leq x} \frac{\log_2 p}{\log_2\sqrt{x}} < \sqrt{x} + \frac{2}{\log_2 x}\sum_{p\leq x}\log_2 p\\ &= \sqrt{x} + \frac{2}{\log_2 x}\log_2\prod_{p\leq x} p \leq \sqrt{x} + \frac{4x}{\log_2 x} \leq \frac{6x}{\log_2 x},\end{aligned}$$

where Lemma 2.2 and the inequality $\sqrt{x} \leq 2x/\log_2 x$ for $x \geq 6$ were used. ∎

Chebyshev in 1848 deduced the estimates with much better constants $a \approx 0.92129$ and $b \approx 1.10555$. To achieve, for example, the better choice of b he used the integers (1.3) from Exercise 1.5 in place of the binomial coefficients $\binom{2m-1}{m}$ as we did in the proof of Lemma 2.2.

2.2 Riemann's zeta function and its basic properties

Riemann's zeta function is a complex-valued function

$$\zeta(s) = \sum_{n=1}^{\infty} \frac{1}{n^s}$$

of argument $s = \sigma + it \in \mathbb{C}$, where $n^s = e^{s\ln n} = n^\sigma n^{it} = n^\sigma(\cos(t\ln n) + i\sin(t\ln n))$ so that $|n^s| = n^\sigma$.

Lemma 2.3. *The series defining the zeta function converges absolutely in the half-plane* $\operatorname{Re} s > 1$ *and defines there the analytic function* $\zeta(s)$. *Furthermore,*

$$\zeta'(s) = -\sum_{n=1}^{\infty} \frac{\ln n}{n^s}.$$

Proof. Fix real $\sigma_0 > 1$. The functions

$$f_n(s) = \frac{1}{n^s}$$

are analytic in the half-plane $D : \operatorname{Re} s \geq \sigma_0$ and the series $\sum_{n=1}^{\infty} f_n(z)$ converges absolutely and uniformly, because of the uniform estimates

$$\sum_{n=1}^{\infty} \left| \frac{1}{n^s} \right| = \sum_{n=1}^{\infty} \frac{1}{n^{\sigma}} \leq \sum_{n=1}^{\infty} \frac{1}{n^{\sigma_0}}$$

valid for all $s \in D$. By the Weierstrass theorem the sum of the series $\zeta(s) = \sum_{n=1}^{\infty} f_n(s)$ is analytic in D and its derivatives $\zeta^{(k)}(s)$ are given by $\sum_{n=1}^{\infty} f_n^{(k)}(s)$ in D for all $k = 1, 2, \dots$.

Because the choice of $\sigma_0 > 1$ is arbitrary, the analyticity of $\zeta(s)$ and representation of $\zeta^{(k)}(s)$ remain valid in the domain $\operatorname{Re} s > 1$. □

The von Mangoldt function is defined for positive integers n by

$$\Lambda(n) = \begin{cases} \ln p & \text{if } n = p^k, \\ 0 & \text{otherwise.} \end{cases}$$

Lemma 2.4. *In the half-plane* $\operatorname{Re} s > 1$, *the representation*

$$-\frac{\zeta'(s)}{\zeta(s)} = \sum_{n=1}^{\infty} \frac{\Lambda(n)}{n^s}$$

is valid.

Proof. As a warm-up we observe that, for $n = p_1^{\alpha_1} \cdots p_m^{\alpha_m}$, we have

$$\begin{aligned} \sum_{d|n} \Lambda(d) &= \sum_{i=1}^{m} \sum_{j=1}^{\alpha_i} \Lambda(p_i^j) = \sum_{i=1}^{m} \sum_{j=1}^{\alpha_i} \ln p_i = \sum_{i=1}^{m} \alpha_i \ln p_i \\ &= \ln(p_1^{\alpha_1} \cdots p_m^{\alpha_m}) = \ln n \end{aligned}$$

(see Exercise 1.1).

Again, let $\sigma_0 > 1$ be fixed. We have

$$\zeta(s) \sum_{k=1}^{\infty} \frac{\Lambda(k)}{k^s} = \sum_{l=1}^{\infty} \frac{1}{l^s} \times \sum_{k=1}^{\infty} \frac{\Lambda(k)}{k^s}$$

(both series converge uniformly in the domain $\operatorname{Re} s \geq \sigma_0$)

$$= \sum_{l=1}^{\infty} \sum_{k=1}^{\infty} \frac{\Lambda(k)}{(lk)^s} = \sum_{n=1}^{\infty} \frac{1}{n^s} \sum_{k|n} \Lambda(k) = \sum_{n=1}^{\infty} \frac{\ln n}{n^s} = -\zeta'(s).$$

The result now follows from noticing that σ_0 can be chosen arbitrarily close to 1. □

Theorem 2.2. *$\zeta(s) \neq 0$ for complex s from the half-plane* $\operatorname{Re} s > 1$.

Proof. Assume this is false and $\zeta(s)$ vanishes at $s = s_0$, $\operatorname{Re} s_0 > 1$. Then the logarithmic derivative of $\zeta(s)$ has a pole of order 1 at this point:

$$\frac{\zeta'(s)}{\zeta(s)} = \frac{C}{s - s_0} + O(1)$$

in a neighbourhood of $s = s_0$; in particular, no limit exists as $s \to s_0$. On the other hand, by Lemma 2.4,

$$\lim_{s \to s_0} \frac{\zeta'(s)}{\zeta(s)} = -\sum_{n=1}^{\infty} \frac{\Lambda(n)}{n^{s_0}},$$

the latter being an absolutely convergent series. Contradiction meaning that no zero of $\zeta(s)$ exists in the half-plane $\operatorname{Re} s > 1$. □

In fact, Riemann's zeta function also does not vanish in an open region that includes the entire line $\operatorname{Re} s = 1$. For our purposes though it will be sufficient to check that $\zeta(s) \neq 0$ on the line, without entering the critical strip $0 < \operatorname{Re} s < 1$.

Exercise 2.1. In the half-plane $\operatorname{Re} s > 1$, prove that

$$\zeta^2(s) = \sum_{n=1}^{\infty} \frac{\tau(n)}{n^s} \quad \text{and} \quad \frac{1}{\zeta(s)} = \sum_{n=1}^{\infty} \frac{\mu(n)}{n^s},$$

where the function $\tau(n)$ is defined in Exercise 1.1 and $\mu(n)$ is the Möbius function.

2.3 Analytic continuation of $\zeta(s)$ to the domain $\operatorname{Re} s > 0$

Lemma 2.5 (Abel transformation). *Let $\{a_k\}_{k=1}^{\infty}$ be a sequence of complex numbers and $g(t)$ a complex-valued differentiable function of real variable $t \in [1, \infty)$. Then*

$$\sum_{1 \leq k \leq x} a_k g(k) = A(x) g(x) - \int_1^x A(t) g'(t) \, dt,$$

where

$$A(t) = \sum_{1 \le k \le t} a_k.$$

Proof. Use the mathematical induction on n, where $n - 1 < x \le n$ (in other words, $n = \lceil x \rceil$). For $n = 1$ we get the obvious identity $a_1 g(1) = A(1)g(1)$. Assume that the required equality is true for all $x \le n$; we will then demonstrate its truth for $x \in (n, n+1]$. Introduce the auxiliary function

$$B(x) = A(x)g(x) - \int_1^x A(t)g'(t)\,\mathrm{d}t,$$

so that we need to show that $B(x) = \sum_{k \le x} a_k g(k)$. We have

$$\begin{aligned} B(x) - B(n) &= A(x)g(x) - A(n)g(n) - \int_n^x A(t)g'(t)\,\mathrm{d}t \\ &= A(x)g(x) - A(n)g(n) - A(n)\int_n^x g'(t)\,\mathrm{d}t \\ &= A(x)g(x) - A(n)g(n) - A(n)(g(x) - g(n)) \\ &= (A(x) - A(n))g(x) = \begin{cases} 0 & \text{if } x < n+1, \\ a_{n+1}g(n+1) & \text{if } x = n+1. \end{cases} \end{aligned}$$

By the inductive hypothesis $B(n) = \sum_{k=1}^n a_k g(k)$ implying the desired formula $B(x) = \sum_{k \le x} a_k g(k)$ for $n < x \le n+1$. □

Consider the Abel transformation in a concrete situation. Take $a_k = 1$ and $g(t) = t^{-s}$, so that $A(x) = \sum_{k \le x} a_k = \lfloor x \rfloor$. Then

$$\begin{aligned} \sum_{n=1}^N \frac{1}{n^s} &= \sum_{n \le N} a_n g(n) = A(N)g(N) + s\int_1^N \frac{\lfloor t \rfloor\,\mathrm{d}t}{t^{s+1}} \\ &= \frac{N}{N^s} + s\int_1^N \frac{\mathrm{d}t}{t^s} - s\int_1^N \frac{\{t\}\,\mathrm{d}t}{t^{s+1}} = \frac{1}{N^{s-1}} + \frac{st^{1-s}}{1-s}\bigg|_{t=1}^N - s\int_1^N \frac{\{t\}\,\mathrm{d}t}{t^{s+1}} \\ &= \frac{1}{N^{s-1}} + \frac{s}{1-s}\frac{1}{N^{s-1}} - \frac{s}{1-s} - s\int_1^N \frac{\{t\}\,\mathrm{d}t}{t^{s+1}} \\ &= \frac{1}{(1-s)N^{s-1}} + \frac{s}{s-1} - s\int_1^N \frac{\{t\}\,\mathrm{d}t}{t^{s+1}}, \end{aligned}$$

where $\{t\}$ is the fractional part. Passing to the limit as $N \to \infty$ we obtain

$$\zeta(s) = \lim_{N \to \infty} \sum_{n=1}^N \frac{1}{n^s} = \frac{s}{s-1} - s\int_1^\infty \frac{\{t\}\,\mathrm{d}t}{t^{s+1}}.$$

To see the correctness of the limiting passage observe that, for each $n \ge 1$, the function

$$h_n(s) = \int_n^{n+1} \frac{\{t\}\,\mathrm{d}t}{t^{s+1}} = \int_n^{n+1} \frac{t-n}{t^{s+1}}\,\mathrm{d}t$$

is analytic and the series

$$\sum_{n=1}^{\infty} h_n(s) = \lim_{N\to\infty} \int_1^{N+1} \frac{\{t\}\,\mathrm{d}t}{t^{s+1}} = \int_1^{\infty} \frac{\{t\}\,\mathrm{d}t}{t^{s+1}}$$

converges absolutely and uniformly in the domain $\operatorname{Re} s \ge \sigma_0$ for any $\sigma_0 > 0$, because of the estimate

$$\left|\frac{\{t\}}{t^{s+1}}\right| \le \frac{1}{t^{\sigma+1}}$$

implying

$$|h_n(s)| \le \int_n^{n+1} \frac{\mathrm{d}t}{t^{\sigma+1}} = \frac{1}{\sigma n^{\sigma}} - \frac{1}{\sigma(n+1)^{\sigma}},$$

hence

$$\sum_{n=1}^{N} |h_n(s)| \le \sum_{n=1}^{N} \left(\frac{1}{\sigma n^{\sigma}} - \frac{1}{\sigma(n+1)^{\sigma}}\right) = \frac{1}{\sigma} - \frac{1}{\sigma(N+1)^{\sigma}} < \frac{1}{\sigma} \le \frac{1}{\sigma_0}$$

for all integers $N \ge 1$. We summarise our finding in the following statement.

Theorem 2.3 (analytic continuation of $\zeta(s)$). *The meromorphic function*

$$\hat{\zeta}(s) = \frac{s}{s-1} - s\int_1^{\infty} \frac{\{t\}\,\mathrm{d}t}{t^{s+1}} = 1 + \frac{1}{s-1} - s\int_1^{\infty} \frac{\{t\}\,\mathrm{d}t}{t^{s+1}}$$

defines the analytic continuation of $\zeta(s)$ to the half-plane $\operatorname{Re} s > 0$, where it has a single pole of order 1 at $s = 1$ with residue 1.

Remark. Another way to analytically continue Riemann's zeta function to the strip $0 < \operatorname{Re} s < 1$ is by means of the representation

$$(1 - 2^{1-s})\zeta(s) = \sum_{n=1}^{\infty} \frac{(-1)^{n-1}}{n^s}.$$

Using it one can show that $\zeta(s)$ does not vanish for *real* s in the range $0 < s < 1$.

2.4 Euler's product and absence of zeros of $\zeta(s)$ on the line $\operatorname{Re} s = 1$

A *multiplicative* function is an arithmetic function $f(n)$ of a positive integer n with the property that $f(1) = 1$ and, whenever a and b are coprime, then $f(ab) = f(a)f(b)$. A function $f(n)$ is said to be *completely* (or *totally*) *multiplicative* if $f(1) = 1$ and $f(ab) = f(a)f(b)$ holds for *all* positive integers a and b.

Lemma 2.6. *Let f be a completely multiplicative function for which the series $S = \sum_{n=1}^{\infty} f(n)$ absolutely converges. Then*

$$S = \prod_p (1 - f(p))^{-1},$$

where the product is over all primes.

Proof. First of all, check that $|f(n)| < 1$ for all $n \geq 2$. Indeed, if this is not the case, $|f(n)| \geq 1$ for some $n \geq 2$, then $|f(n^k)| = |f(n)|^k \geq 1$ so that the necessary condition $f(n^k) \to 0$ as $k \to \infty$ for convergence of the series S is violated. With the bound $|f(p)| < 1$ for any prime p in mind, we can write the geometric series

$$(1 - f(p))^{-1} = \sum_{k=0}^{\infty} f(p)^k = \sum_{k=0}^{\infty} f(p^k),$$

which together with the complete multiplicativity of $f(n)$ and the fundamental theorem of arithmetic (Theorem 1.1) imply

$$\prod_{p \leq x} (1 - f(p))^{-1} = \sum_{\substack{n = p_1^{\alpha_1} \cdots p_m^{\alpha_m} \\ p_i \leq x}} f(n) = S - \sum_{n : p \mid n \text{ for some } p > x} f(n).$$

It follows from this result that

$$\left| S - \prod_{p \leq x} (1 - f(p))^{-1} \right| \leq \sum_{n : p \mid n \text{ for some } p > x} |f(n)| \leq \sum_{n > x} |f(n)| = \sum_{n > \lfloor x \rfloor} |f(n)|.$$

The latter sum is a tail of the absolutely convergent series $\sum_{n=1}^{\infty} |f(n)|$, thus it tends to 0 as $x \to \infty$. This proves the required limiting relation. ☐

Theorem 2.4 (Euler's product for $\zeta(s)$). *In the half-plane* $\operatorname{Re} s > 1$, *the following representation takes place*:

$$\zeta(s) = \prod_p \left(1 - \frac{1}{p^s} \right)^{-1}.$$

Proof. Apply Lemma 2.6 to the completely multiplicative function $f(n) = 1/n^s$. □

Exercise 2.2 (Euler). Use Theorem 2.4 to show that the series

$$\sum_p \frac{1}{p}$$

diverges. Conclude from this that there are infinitely many primes.

Now notice the following nice but elementary inequality.

Lemma 2.7. *We have*

$$|(1-r)^3(1-re^{i\theta})^4(1-re^{2i\theta})| \le 1, \quad \textit{where } 0 < r < 1.$$

Proof. Denote $M = |(1-r)^3(1-re^{i\theta})^4(1-re^{2i\theta})|$ and observe that $\operatorname{Re}\ln(1-z) = \ln|1-z|$ for all z inside the unit disc, $|z| < 1$. Therefore,

$$\begin{aligned}
\ln M &= 3\ln|1-r| + 4\ln|1-re^{i\theta}| + \ln|1-re^{2i\theta}| \\
&= \operatorname{Re}\big(3\ln(1-r) + 4\ln(1-re^{i\theta}) + \ln(1-re^{2i\theta})\big) \\
&= -\operatorname{Re}\left(3\sum_{n=1}^{\infty}\frac{r^n}{n} + 4\sum_{n=1}^{\infty}\frac{r^n e^{in\theta}}{n} + \sum_{n=1}^{\infty}\frac{r^n e^{2in\theta}}{n}\right) \\
&= -\sum_{n=1}^{\infty}\frac{r^n}{n}\operatorname{Re}(3 + 4e^{in\theta} + e^{2in\theta}) \\
&= -\sum_{n=1}^{\infty}\frac{r^n}{n}(3 + 4\cos n\theta + \cos 2n\theta) \\
&= -\sum_{n=1}^{\infty}\frac{r^n}{n} \times 2(1+\cos n\theta)^2 \le 0
\end{aligned}$$

implying that $M \le 1$. □

Theorem 2.5. *If* $\operatorname{Re} s = 1$ *then* $\zeta(s) \ne 0$.

Proof. It follows from the lemma and Euler's representation of Riemann's zeta function (Theorem 2.4) that

$$|\zeta^3(\sigma)\zeta^4(\sigma+it)\zeta(\sigma+2it)| = \prod_p |(1-p^{-\sigma})^3(1-p^{-(\sigma+it)})^4(1-p^{-(\sigma+2it)})|^{-1} \ge 1$$

for $\sigma > 1$. Furthermore, Theorem 2.3 implies that

$$\zeta(\sigma) = \frac{\sigma}{\sigma-1} - \sigma\int_1^{\infty}\frac{\{t\}\,\mathrm{d}t}{t^{\sigma+1}} \le \frac{\sigma}{\sigma-1},$$

so that $\zeta(\sigma) = O((\sigma-1)^{-1})$ as $\sigma \to 1^+$. Assume that $s_0 = 1 + it_0$ for some $t_0 \neq 0$ is a zero of $\zeta(s)$ on the line $\operatorname{Re} s = 1$. Then $\zeta(\sigma + it_0) = O(s - s_0) = O(\sigma - 1)$ and $\zeta(\sigma + 2it_0) = O(1)$ as $\sigma \to 1^+$. Then

$$\begin{aligned}|\zeta^3(\sigma)\zeta^4(\sigma+it_0)\zeta(\sigma+2it_0)| &= O\big((\sigma-1)^{-3}\cdot(\sigma-1)^4\cdot 1\big)\\ &= O(\sigma-1) \quad \text{as } \sigma \to 1^+,\end{aligned}$$

so that $|\zeta^3(\sigma)\zeta^4(\sigma+it_0)\zeta(\sigma+2it_0)|$ can be made arbitrary close to 0 contradicting to the earlier established bound for the expression. □

2.5 Upper estimates for $\zeta'(s)/\zeta(s)$

In what follows $s = \sigma + it$. The goal of this section is to give upper estimates for the absolute value of the logarithmic derivative of $\zeta(s)$ in the domain $1 \le \sigma \le 2$, $|t| \ge 3$.

Lemma 2.8. *In the domain* $1 \le \sigma \le 2$, $|t| \ge 3$, *the following estimates take place*:

$$|\zeta(s)| \le 5\ln|t| \quad \textit{and} \quad |\zeta'(s)| \le 8\ln^2|t|.$$

Proof. In the domain under consideration, the function $\zeta(s)$ is analytic and computed by the formula

$$\zeta(s) = \sum_{n=1}^{N} \frac{1}{n^s} + \frac{1}{(s-1)N^{s-1}} - s\int_N^\infty \frac{\{t\}\,dt}{t^{s+1}}$$

(see Section 2.3). Differentiating both sides of the representation in the domain we also get

$$\begin{aligned}\zeta'(s) = &-\sum_{n=1}^{N} \frac{\ln n}{n^s} - \frac{1}{(s-1)^2N^{s-1}} - \frac{\ln N}{(s-1)N^{s-1}}\\ &-\int_N^\infty \frac{\{t\}\,dt}{t^{s+1}} + s\int_N^\infty \frac{\{t\}\ln t\,dt}{t^{s+1}}.\end{aligned}$$

In these formulas we choose $N = \lfloor |t| \rfloor \ge 3$. Then

$$\begin{aligned}\left|\sum_{n=1}^{N} \frac{1}{n^s}\right| &\le \sum_{n=1}^{N} \left|\frac{1}{n^s}\right| = \sum_{n=1}^{N} \frac{1}{n^\sigma} \le \sum_{n=1}^{N} \frac{1}{n} \le 1 + \int_1^N \frac{dx}{x}\\ &= 1 + \ln N \le 2\ln|t|,\end{aligned}$$

$$\begin{aligned}\left|\sum_{n=1}^{N} \frac{\ln n}{n^s}\right| &\le \sum_{n=1}^{N} \frac{\ln n}{n} \le \frac{\ln 2}{2} + \frac{\ln 3}{3} + \int_3^N \frac{\ln x\,dx}{x}\\ &= \frac{\ln 2}{2} + \frac{\ln 3}{3} + \frac{\ln^2 N}{2} - \frac{\ln^2 3}{2} \le \ln^2 N \le \ln^2|t|,\end{aligned}$$

$$\left|\int_N^\infty \frac{\{t\}\,\mathrm{d}t}{t^{s+1}}\right| \le \int_N^\infty \frac{\mathrm{d}t}{t^{\sigma+1}} \le \int_N^\infty \frac{\mathrm{d}t}{t^2} = \frac{1}{N} \le 1,$$
$$\left|s\int_N^\infty \frac{\{t\}\,\mathrm{d}t}{t^{s+1}}\right| \le \frac{|s|}{N} \le \frac{\sigma+|t|}{N} \le \frac{2+(N+1)}{N} = 1+\frac{3}{N} \le 2,$$
$$\left|\frac{1}{(s-1)N^{s-1}}\right| = \frac{1}{|s-1|\,N^{\sigma-1}} \le \frac{1}{|s-1|} \le 1,$$
$$\left|\frac{1}{(s-1)^2N^{s-1}}\right| \le \frac{1}{|s-1|^2} \le 1,$$
$$\left|\frac{\ln N}{(s-1)N^{s-1}}\right| \le \frac{\ln N}{|s-1|} \le \ln N \le \ln|t|$$

and, finally,

$$\left|s\int_N^\infty \frac{\{t\}\ln t\,\mathrm{d}t}{t^{s+1}}\right| \le |s|\int_N^\infty \frac{\ln t\,\mathrm{d}t}{t^2} = |s|\left(-\frac{1+\ln t}{t}\right)\Bigg|_{t=N}^\infty$$
$$= \frac{|s|}{N}(1+\ln N) \le 2(1+\ln|t|).$$

Thus,

$$|\zeta(s)| \le 2\ln|t| + 1 + 2 \le 5\ln|t|,$$
$$|\zeta'(s)| \le \ln^2|t| + 1 + \ln|t| + 1 + 2(1+\ln|t|) \le 8\ln^2|t|. \qquad \square$$

Lemma 2.9. *In the domain* $1 \le \sigma \le 2$, $|t| \ge 3$, *the following estimate holds*:

$$\left|\frac{\zeta'(s)}{\zeta(s)}\right| \le C\ln^9|t|, \quad \textit{where } C = 2^{23}.$$

Proof. By Lemma 2.7 and Theorem 2.4 we have

$$|\zeta^3(\sigma)\zeta^4(\sigma+it)\zeta(\sigma+2it)| \ge 1$$

(see also the proof of Theorem 2.5). In particular, we conclude that

$$|\zeta(s)| = |\zeta(\sigma+it)| \ge |\zeta(\sigma)|^{-3/4}|\zeta(\sigma+2it)|^{-1/4}.$$

It follows from Theorem 2.3 that

$$\zeta(\sigma) = \frac{\sigma}{\sigma-1} - \sigma\int_1^\infty \frac{\{t\}\,\mathrm{d}t}{t^{\sigma+1}} \le \frac{\sigma}{\sigma-1}$$

for all $\sigma > 1$, so that

$$\zeta(\sigma) \le \frac{2}{\sigma-1} \le 2C\ln^9|t|$$

for all $\sigma \geq \sigma_1 = 1 + 1/(C\ln^9|t|)$. Lemma 2.8 implies the estimate

$$|\zeta(\sigma + 2it)| \leq 5\ln(2|t|) \leq 16\ln|t|,$$

so that

$$|\zeta(s)| \geq (2C\ln^9|t|)^{-3/4}(16\ln|t|)^{-1/4} = 16C^{-1}\ln^{-7}|t|$$

in the domain $\sigma_1 \leq \sigma \leq 2$, $|t| \geq 3$. But then for σ in the range $1 \leq \sigma < \sigma_1$ we have from the mean value theorem and the estimate of Lemma 2.8 for $\zeta'(s)$,

$$\begin{aligned}
|\zeta(s)| &\geq |\zeta(\sigma_1 + it)| - |\zeta(\sigma_1 + it) - \zeta(\sigma + it)| \\
&= |\zeta(\sigma_1 + it)| - \left|\int_{\sigma}^{\sigma_1} \zeta'(u + it)\,du\right| \\
&\geq 16C^{-1}\ln^{-7}|t| - (\sigma_1 - \sigma) \times 8\ln^2|t| \\
&\geq 16C^{-1}\ln^{-7}|t| - 8C^{-1}\ln^{-7}|t| \\
&= 8C^{-1}\ln^{-7}|t|.
\end{aligned}$$

This means that the estimate

$$|\zeta(s)| \geq 8C^{-1}\ln^{-7}|t|$$

holds for all s from the domain $1 \leq \sigma \leq 2$, $|t| \geq 3$. Then Lemma 2.8 is used again to conclude with the estimate

$$\left|\frac{\zeta'(s)}{\zeta(s)}\right| \leq \frac{8\ln^2|t|}{8C^{-1}\ln^{-7}|t|} = C\ln^9|t|. \qquad \square$$

2.6 Chebyshev's function $\psi(x)$. Reduction of the prime number theorem

Recall the prime counting function

$$\pi(x) = \sum_{p \leq x} 1$$

and introduce the related Chebyshev function

$$\psi(x) = \sum_{n \leq x} \Lambda(n) = \sum_{p^m \leq x} \ln p,$$

where the summation is over all pairs of primes p and exponents m subject to $p^m \leq x$. The latter inequality implies that $m \leq (\ln x)/(\ln p)$ so that

$$\psi(x) = \sum_{p \leq x} \left\lfloor \frac{\ln x}{\ln p} \right\rfloor \ln p.$$

In particular, we have

$$\pi(x)\ln x - \psi(x) = \sum_{p \leq x} \left(\frac{\ln x}{\ln p} - \left\lfloor \frac{\ln x}{\ln p} \right\rfloor\right) \ln p = \sum_{p \leq x} \left\{\frac{\ln x}{\ln p}\right\} \ln p \geq 0$$

for all $x > 0$.

Lemma 2.10. *The following asymptotic takes place:* $\psi(x) = \pi(x)\ln x + o(x)$ *as* $x \to \infty$.

Proof. Clearly,

$$\left\{\frac{\ln x}{\ln p}\right\} \ln p \le \ln p$$

for all primes p, and we also have

$$\left\{\frac{\ln x}{\ln p}\right\} \ln p = \ln x - \left\lfloor \frac{\ln x}{\ln p} \right\rfloor \ln p \le \ln x - \ln p = \ln \frac{x}{p}$$

when $p \le x$. Using now the upper bound $\pi(y) < by/(\ln y)$ for $y \ge 2$ from Theorem 2.1 we find, for $x \ge 8 > e^2$, that

$$\begin{aligned}
\pi(x)\ln x - \psi(x) &= \sum_{p \le x/(\ln x)} \left\{\frac{\ln x}{\ln p}\right\} \ln p + \sum_{x/(\ln x) < p \le x} \left\{\frac{\ln x}{\ln p}\right\} \ln p \\
&\le \sum_{p \le x/(\ln x)} \ln p + \sum_{x/(\ln x) < p \le x} \ln \frac{x}{p} \\
&\le \sum_{p \le x/(\ln x)} \ln\left(\frac{x}{\ln x}\right) + \sum_{x/(\ln x) < p \le x} \ln(\ln x) \\
&\le \ln\left(\frac{x}{\ln x}\right) \cdot \pi\left(\frac{x}{\ln x}\right) + \ln\ln x \cdot \pi(x) \\
&\le b\frac{x}{\ln x} + \ln\ln x \cdot \frac{bx}{\log x} = \frac{bx(1 + \ln\ln x)}{\ln x}.
\end{aligned}$$

Thus,

$$0 \le \frac{\pi(x)\ln x - \psi(x)}{x} \le \frac{b(1 + \ln\ln x)}{\ln x} \to 0 \quad \text{as } x \to \infty$$

and the required asymptotics follows. □

Lemma 2.10 means that the asymptotic distribution of primes,

$$\pi(x) \sim \frac{x}{\ln x} \quad \text{as } x \to \infty,$$

is equivalent to $\psi(x) = x + o(x)$ as $x \to \infty$. The next statement reduces establishing of the latter to verifying the asymptotic relation $\omega(x) = x + o(x)$ as $x \to \infty$, where

$$\omega(x) = \int_1^x \frac{\psi(t)}{t}\,\mathrm{d}t.$$

Lemma 2.11 (further reduction). *If* $\omega(x) = x + o(x)$ *as* $x \to \infty$, *then also* $\psi(x) = x + o(x)$ *and so* $\pi(x) \sim x/(\ln x)$ *as* $x \to \infty$.

Proof. Suppose that the asymptotics $\omega(x) = x + o(x)$ as $x \to \infty$ is established. Take an arbitrary ε in the range $0 < \varepsilon < 1$. Because the function $\psi(t)$ is monotone increasing, we get

$$\omega((1+\varepsilon)x) - \omega(x) = \int_x^{(1+\varepsilon)x} \frac{\psi(t)\,\mathrm{d}t}{t} \geq \psi(x) \int_x^{(1+\varepsilon)x} \frac{\mathrm{d}t}{t} = \psi(x)\ln(1+\varepsilon),$$

hence

$$\limsup_{x\to\infty} \frac{\psi(x)}{x} \leq \frac{1}{\ln(1+\varepsilon)} \lim_{x\to\infty} \frac{\omega((1+\varepsilon)x) - \omega(x)}{x} = \frac{\varepsilon}{\ln(1+\varepsilon)}.$$

Since the estimate is true for any $\varepsilon > 0$, it remains valid as $\varepsilon \to 0^+$ and we obtain

$$\limsup_{x\to\infty} \frac{\psi(x)}{x} \leq \lim_{\varepsilon\to 0^+} \frac{\varepsilon}{\ln(1+\varepsilon)} = 1.$$

Similar consideration leads to

$$\omega(x) - \omega((1-\varepsilon)x) = \int_{(1-\varepsilon)x}^{x} \frac{\psi(t)\,\mathrm{d}t}{t} \leq \psi(x) \int_{(1-\varepsilon)x}^{x} \frac{\mathrm{d}t}{t} = \psi(x) \ln\frac{1}{1-\varepsilon},$$

therefore

$$\liminf_{x\to\infty} \frac{\psi(x)}{x} \geq \frac{1}{\ln(1/(1-\varepsilon))} \lim_{x\to\infty} \frac{\omega(x) - \omega((1-\varepsilon)x)}{x} = \frac{\varepsilon}{-\ln(1-\varepsilon)}$$

and

$$\liminf_{x\to\infty} \frac{\psi(x)}{x} \geq \lim_{\varepsilon\to 0^+} \frac{\varepsilon}{-\ln(1-\varepsilon)} = 1.$$

□

2.7 Integral representation for $\omega(x)$

Lemma 2.12. *For $a, b > 0$, we have*

$$\frac{1}{2\pi i} \int_{a-i\infty}^{a+i\infty} \frac{b^s}{s^2}\,\mathrm{d}s = \begin{cases} \ln b & \text{if } b \geq 1, \\ 0 & \text{if } 0 < b < 1, \end{cases}$$

where the integration is performed along the vertical line $\operatorname{Re} s = a$.

Proof. For $s = a + it$,

$$\left|\frac{b^s}{s^2}\right| = \frac{b^a}{a^2 + t^2};$$

therefore, the integral under consideration converges absolutely. Consider first the case $b \geq 1$ and the integral

$$I(r) = \frac{1}{2\pi i} \int_\Gamma \frac{b^s}{s^2}\,\mathrm{d}s$$

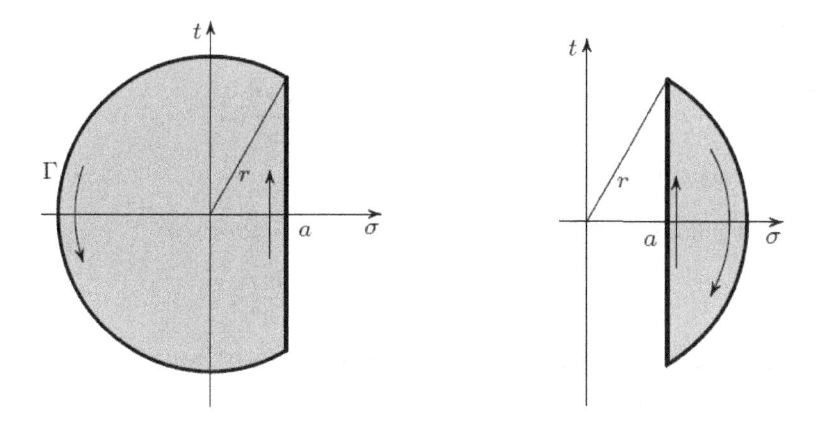

Fig. 2.1 The contour for $b \geq 1$ (left) and $0 < b < 1$ (right)

along the closed contour (on the left of Fig. 2.1) consisting of the arc of radius r centred at 0 from the left of the line $\operatorname{Re} s = a$ and the segment joining the endpoints of the arc on the line.

Inside the contour, the integrand has a single pole of order 2 at $s = 0$ and, because of the Laurent expansion

$$\frac{b^s}{s^2} = \frac{e^{s \ln b}}{s^2} = \frac{1}{s^2} + \frac{\ln b}{s} + \frac{1}{2} \ln^2 b + \cdots ,$$

we have $I(r) = \ln b$ so that

$$\frac{1}{2\pi i} \int_{a-ir_0}^{a+ir_0} \frac{b^s}{s^2} \, \mathrm{d}s = \ln b - \frac{1}{2\pi i} \int_{\text{arc}} \frac{b^s}{s^2} \, \mathrm{d}s,$$

where $r_0 = \sqrt{r^2 - a^2}$. Furthermore, $|b^s| = b^{\operatorname{Re} s} \leq b^a$ on the contour, since $b \geq 1$ and $\operatorname{Re} s \leq a$, and for the integral along the arc alone,

$$\left| \frac{1}{2\pi i} \int_{\text{arc}} \frac{b^s}{s^2} \, \mathrm{d}s \right| \leq \frac{1}{2\pi} \int_{\text{arc}} |\mathrm{d}s| \times \frac{b^a}{r^2} \leq \frac{b^a}{r} \to 0 \quad \text{as } r \to \infty,$$

implying that

$$\frac{1}{2\pi i} \lim_{r_0 \to \infty} \int_{a-ir_0}^{a+ir_0} \frac{b^s}{s^2} \, \mathrm{d}s = \ln b.$$

In the case $0 < b < 1$, we use the closed contour Γ depicted on the right of Fig. 2.1. The corresponding integral $I(r)$ now vanishes, because there are no poles inside Γ. In this case $|b^s| = b^{\operatorname{Re} s} \leq b^a$ on Γ, so that the same estimate for the integral along the arc is valid, and we conclude with the value as above. □

Theorem 2.6. *For $a > 1$ and $x \geq 2$, the following integral representation is valid:*

$$\omega(x) = \frac{1}{2\pi i}\int_{a-i\infty}^{a+i\infty}\left(-\frac{\zeta'(s)}{\zeta(s)}\right)\frac{x^s}{s^2}\,\mathrm{d}s,$$

and the integral absolutely converges.

Proof. It follows from Lemma 2.4 that on the line $\mathrm{Re}\, s = a$,

$$\left|-\frac{\zeta'(s)}{\zeta(s)}\right| \leq \sum_{n=2}^{\infty}\frac{\Lambda(n)}{n^a} \leq \sum_{n=2}^{\infty}\frac{\ln n}{n^a}.$$

This implies the absolute convergence of the integral. In the Abel transformation, Lemma 2.5, take $a_k = \Lambda(k)$ and $g(t) = \ln(x/t)$. Then $A(t) = \sum_{k\leq t} a_k = \psi(t)$, Chebyshev's function, hence

$$\sum_{n\leq x}\Lambda(n)\ln\frac{x}{n} = \psi(x)\ln 1 + \int_1^x \psi(t)\,\frac{\mathrm{d}t}{t} = \omega(x).$$

It follows from Lemma 2.12 that

$$\frac{1}{2\pi i}\int_{a-i\infty}^{a+i\infty}\left(\frac{x}{n}\right)^s\frac{\mathrm{d}s}{s^2} = \begin{cases}\ln(x/n) & \text{if } n \leq x,\\ 0 & \text{if } n > x,\end{cases}$$

while Lemma 2.4 implies

$$\left(-\frac{\zeta'(s)}{\zeta(s)}\right)\frac{x^s}{s^2} = \sum_{n=2}^{\infty}\Lambda(n)\left(\frac{x}{n}\right)^s\frac{1}{s^2}.$$

Combining the two results we obtain

$$\begin{aligned}\omega(x) &= \sum_{n\leq x}\Lambda(n)\ln\frac{x}{n} = \sum_{n\leq x}\Lambda(n)\,\frac{1}{2\pi i}\int_{a-i\infty}^{a+i\infty}\left(\frac{x}{n}\right)^s\frac{\mathrm{d}s}{s^2}\\ &= \sum_{n=2}^{\infty}\Lambda(n)\,\frac{1}{2\pi i}\int_{a-i\infty}^{a+i\infty}\left(\frac{x}{n}\right)^s\frac{\mathrm{d}s}{s^2}\\ &= \frac{1}{2\pi i}\int_{a-i\infty}^{a+i\infty}\sum_{n=2}^{\infty}\Lambda(n)\left(\frac{x}{n}\right)^s\frac{\mathrm{d}s}{s^2}\\ &= \frac{1}{2\pi i}\int_{a-i\infty}^{a+i\infty}\left(-\frac{\zeta'(s)}{\zeta(s)}\right)\frac{x^s}{s^2}\,\mathrm{d}s,\end{aligned}$$

the desired identity, so we only need to justify the interchange of summation and integration. For the latter, notice that

$$\left|\Lambda(n)\left(\frac{x}{n}\right)^s\frac{1}{s^2}\right| \leq \frac{\ln n}{a^2}\left(\frac{x}{n}\right)^a$$

on the contour of integration $\operatorname{Re} s = a$, hence the series

$$\sum_{n=2}^{\infty} \Lambda(n) \left(\frac{x}{n}\right)^s \frac{1}{s^2}$$

converges absolutely and uniformly on the line. This means that for each $T > 0$, we have the equality

$$\frac{1}{2\pi i} \int_{a-iT}^{a+iT} \left(-\frac{\zeta'(s)}{\zeta(s)}\right) \frac{x^s}{s^2}\, \mathrm{d}s = \sum_{n=2}^{\infty} \frac{1}{2\pi i} \int_{a-iT}^{a+iT} \Lambda(n) \left(\frac{x}{n}\right)^s \frac{\mathrm{d}s}{s^2},$$

while the estimate

$$\left| \frac{1}{2\pi i} \int_{a-iT}^{a+iT} \Lambda(n) \left(\frac{x}{n}\right)^s \frac{\mathrm{d}s}{s^2} \right| \le \frac{\ln n}{2\pi} \left(\frac{x}{n}\right)^a \int_{-\infty}^{\infty} \frac{\mathrm{d}t}{a^2 + t^2} = \frac{\ln n}{2a} \left(\frac{x}{n}\right)^a$$

implies that the series over n converges uniformly on the set $T > 0$, so the transition as $T \to \infty$ is legal. This completes the proof of the theorem. □

2.8 The principal asymptotics of $\omega(x)$

Theorem 2.7. *Suppose that $\zeta(s)$ does not vanish for $s = \sigma + it$ inside the closed rectangle $\eta \le \sigma \le 1$, $|t| \le T$ for some $\eta < 1$ and $T > 0$. Then $\omega(x) = (1 + R(x))x$, where*

$$R(x) = \frac{1}{2\pi i} \int_{\Gamma(T,\eta)} \left(-\frac{\zeta'(s)}{\zeta(s)}\right) \frac{x^{s-1}}{s^2}\, \mathrm{d}s$$

for the contour $\Gamma(T,\eta)$ depicted on the left of Fig. 2.2.

Proof. For $u > T$, consider the contour $\Gamma = \Gamma(u, T, \eta)$ on the right of Fig. 2.2, which is symmetric along the real axis and in which the imaginary parts of points B, C are equal to u. The function $\zeta(s)$ does not vanish inside the contour and has a single simple point with residue 1 at $s = 1$, so that $\zeta(s) = 1/(s-1) + f(s)$ for some $f(s)$ analytic inside and on the boundary of Γ. Then

$$-\frac{\zeta'(s)}{\zeta(s)} = \frac{1 - (s-1)^2 f'(s)}{1 + (s-1)f(s)} \times \frac{1}{s-1},$$

hence the function

$$\left(-\frac{\zeta'(s)}{\zeta(s)}\right) \frac{x^s}{s^2}$$

has a single singularity inside Γ—the simple pole at $s = 1$ with residue $x^s|_{s=1} = x$ implying

$$\frac{1}{2\pi i} \int_{\Gamma(u,T,\eta)} \left(-\frac{\zeta'(s)}{\zeta(s)}\right) \frac{x^s}{s^2}\, \mathrm{d}s = x.$$

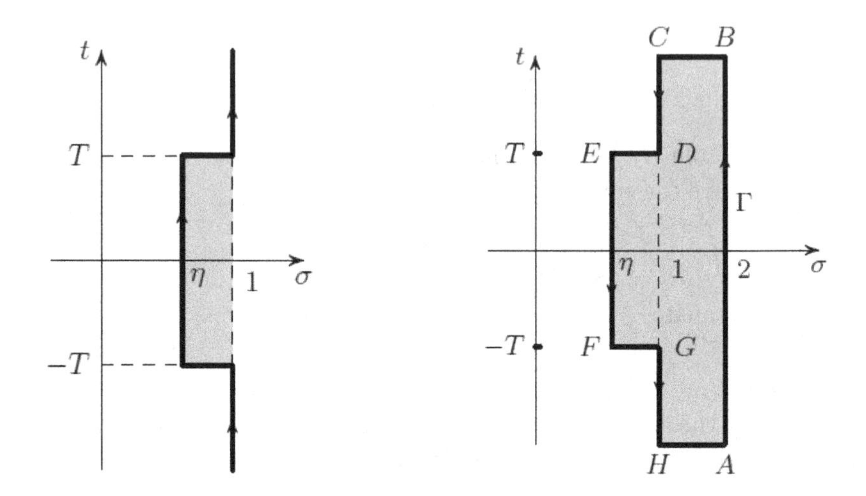

Fig. 2.2 The contour in Theorem 2.7 (left) and in its proof (right)

Now estimate the integral along the segment BC (and similarly, along HA):

$$\left|\frac{1}{2\pi i}\int_{1+iu}^{2+iu}\left(-\frac{\zeta'(s)}{\zeta(s)}\right)\frac{x^{s-1}}{s^2}\,\mathrm{d}s\right| \le C\frac{\ln^9 u}{u^2}x^2 \to 0 \quad \text{as } u \to \infty.$$

Thus, taking the limit as $u \to \infty$ in the former expression and using

$$\omega(x) = \frac{1}{2\pi i}\int_{2-i\infty}^{2+i\infty}\left(-\frac{\zeta'(s)}{\zeta(s)}\right)\frac{x^s}{s^2}\,\mathrm{d}s$$

(here the convergence is absolute), we arrive at the desired claim. □

Lemma 2.13. *For the function $R(x)$ defined in Theorem 2.7 we have $R(x) \to 0$ as $x \to \infty$.*

Proof. Notice that we can manipulate with choosing $\eta < 1$ and $T > 0$; the only constraints is that $\zeta(s)$ should not vanish inside and on the sides of the rectangle $\eta \le \operatorname{Re} s \le 1$, $|\operatorname{Im} s| \le T$.

Take an arbitrary $\varepsilon > 0$. It follows from Lemma 2.9 that

$$\left|\frac{\zeta'(1+it)}{\zeta(1+it)}\frac{x^{it}}{(1+it)^2}\right| \le \frac{C\ln^9|t|}{1+t^2} \quad \text{for } |t| \ge 3.$$

This means that we can pick up some $T = T(\varepsilon) > 3$ independent of x, for which

$$\left|\frac{1}{2\pi i}\int_{1+iT}^{1+i\infty}\left(-\frac{\zeta'(s)}{\zeta(s)}\right)\frac{x^{s-1}}{s^2}\,\mathrm{d}s\right| \le \frac{1}{2\pi}\int_T^{\infty}\frac{C\ln^9|t|}{1+t^2}\,\mathrm{d}t < \frac{\varepsilon}{5}$$

and

$$\left|\frac{1}{2\pi i}\int_{1-i\infty}^{1-iT}\left(-\frac{\zeta'(s)}{\zeta(s)}\right)\frac{x^{s-1}}{s^2}\,\mathrm{d}s\right| \le \frac{1}{2\pi}\int_T^\infty \frac{C\ln^9|t|}{1+t^2}\,\mathrm{d}t < \frac{\varepsilon}{5}.$$

Since the function $\zeta(s)$ does not vanish on the interval $[1-iT, 1+iT]$ for T so chosen, we can choose some $\delta < 1$ in such a way that there are no zeros of $\zeta(s)$ inside the rectangle $\eta \le \operatorname{Re} s \le 1$, $|\operatorname{Im} s| \le T$. The function

$$\left(-\frac{\zeta'(s)}{\zeta(s)}\right)\frac{1}{s^2}$$

is continuous on the sides $[1-iT, \eta-iT]$, $[\eta-iT, \eta+iT]$ and $[\eta+iT, 1+iT]$ of the rectangle, hence

$$\left|\left(-\frac{\zeta'(s)}{\zeta(s)}\right)\frac{1}{s^2}\right| \le M$$

on those sides for some $M = M(T,\eta) > 0$. This implies that

$$\begin{aligned}\left|\frac{1}{2\pi i}\int_{\eta+iT}^{1+iT}\left(-\frac{\zeta'(s)}{\zeta(s)}\right)\frac{x^{s-1}}{s^2}\,\mathrm{d}s\right| &\le \frac{M}{2\pi}\int_\eta^1 x^{\sigma-1}\,\mathrm{d}\sigma = \frac{M}{2\pi x}\int_\eta^1 e^{\sigma\ln x}\,\mathrm{d}\sigma \\ &= \frac{M}{2\pi x\ln x}\,e^{\sigma\ln x}\Big|_{\sigma=\eta}^1 \\ &< \frac{M}{2\pi x\ln x}\,e^{\ln x} = \frac{M}{2\pi\ln x}\end{aligned}$$

and, similarly,

$$\left|\frac{1}{2\pi i}\int_{1-iT}^{\eta-iT}\left(-\frac{\zeta'(s)}{\zeta(s)}\right)\frac{x^{s-1}}{s^2}\,\mathrm{d}s\right| < \frac{M}{2\pi\ln x},$$

while

$$\left|\frac{1}{2\pi i}\int_{\eta-iT}^{\eta+iT}\left(-\frac{\zeta'(s)}{\zeta(s)}\right)\frac{x^{s-1}}{s^2}\,\mathrm{d}s\right| \le \frac{Mx^{\sigma-1}}{2\pi}\int_{-T}^T \mathrm{d}t = \frac{MT}{\pi}x^{\sigma-1}.$$

By choosing x sufficiently large, $x > X = X(T,\eta)$, we can make all these latter three integrals less than $\varepsilon/5$. Combining the five estimates for the integrals involved in computation of $R(x)$, we see that there is a choice of $T > 0$, $\eta < 1$ and $X > 0$ such that

$$|R(x)| = \left|\frac{1}{2\pi i}\int_{\Gamma(T,\eta)}\left(-\frac{\zeta'(s)}{\zeta(s)}\right)\frac{x^{s-1}}{s^2}\,\mathrm{d}s\right| \le \varepsilon$$

for all $x > X$. This concludes the proof of our lemma. □

Theorem 2.8 (prime number theorem). *For the prime counting function $\pi(x)$, we have*

$$\pi(x) \sim \frac{x}{\ln x} \quad \textit{as } x \to \infty.$$

Proof. It follows from Theorem 2.7 and Lemma 2.13 that $\omega(x) = x + o(x)$ as $x \to \infty$. By Lemma 2.11 this implies the asymptotics of $\pi(x)$. □

Exercise 2.3. Let $p_1 = 2 < p_2 = 3 < \cdots < p_n < \cdots$ be the sequence of all primes. Show that

$$p_n \sim n \ln n \quad \text{as } n \to \infty.$$

Exercise 2.4. For $\operatorname{Re} s > 1$, show that

$$\Gamma(s)\zeta(s) = \int_0^{+\infty} \frac{x^{s-1}}{e^x - 1}\, \mathrm{d}x.$$

Exercise 2.5. Show that if $a > 1$ and $x > 1$ is not an integer then

$$\sum_{1 \le n \le x} \mu(n) = \frac{1}{2\pi i} \int_{a-i\infty}^{a+i\infty} \frac{1}{\zeta(s)} \frac{x^s}{s}\, \mathrm{d}s,$$

where $\mu(n)$ is the Möbius function.

Hint. Use Exercise 2.1. □

Exercise 2.6. Use the previous exercise to deduce that

$$\sum_{1 \le n \le x} \mu(n) = o(x) \quad \text{as } x \to \infty.$$

Chapter notes

Based on Euler's ideas (1737), Riemann's memoir *Ueber die Anzahl der Primzahlen unter einer gegebenen Grösse* (1859) set up an analytic approach to questions concerning the distribution of prime numbers, in particular connecting those with the zeros of $\zeta(s)$ as a function of complex variable s. Realisations of Riemann's ideas for the asymptotic law of the distribution of primes were found independently and in the same year (1896) by J. Hadamard and Ch. J. de la Vallée Poussin (in Chapter 8 we witness another longstanding problem with independent and simultaneous resolution — such coincidences are common in number theory). Decades later different proofs of the prime number theorem were found including the 'elementary' proofs (without use of complex analysis, at the cost of being more manipulative) of A. Selberg and P. Erdős (both published in 1949, with independence of Erdős's proof disputed).

It is generally recognised that problems about prime numbers represent very old and most challenging problems not only in number theory but in mathematics in general; for this reason *Analytic Number Theory* is often understood as a field dedicated exclusively to those. A spectacular recent progress is already hard to overview, so we limit ourselves to mentioning the Green–Tao theorem about arbitrarily long arithmetic progressions of prime numbers [37] and the infinitude of bounded gaps between primes proven by Y. Zhang [86] and later by J. Maynard [57] using a different method.

We return to the 'prime' topic in Chapter 5 to discuss the infinitude of prime numbers in arithmetic progressions, the result proven by Dirichlet long before Riemann's memoir.

Chapter 3

Riemann's zeta function and its multiple generalisation

3.1 Euler's gamma function

Riemann's zeta function is always accompanied by Euler's gamma function $\Gamma(z)$ defined through the product expansion

$$\frac{1}{\Gamma(z)} = ze^{\gamma z} \prod_{k=1}^{\infty} \left(1 + \frac{z}{k}\right) e^{-z/k} \tag{3.1}$$

for its reciprocal. Here

$$\begin{aligned} \gamma &= \lim_{n\to\infty} \left(1 + \frac{1}{2} + \frac{1}{3} + \cdots + \frac{1}{n} - \log n\right) \\ &= 0.57721566490153286060651209008240243104215933593992\ldots \end{aligned}$$

is the *Euler* (or *Euler–Mascheroni*) *constant.* A theorem of Weierstrass guarantees that $1/\Gamma(z)$ is an entire function with zeros at $z = 0, -1, -2, \ldots,$ and many properties of the gamma function, like the difference equation

$$\Gamma(z+1) = z\Gamma(z), \tag{3.2}$$

the reflection formula

$$\Gamma(z)\Gamma(1-z) = \frac{1}{z} \prod_{k=1}^{\infty} \left(1 - \frac{z^2}{k^2}\right)^{-1} = \frac{\pi}{\sin \pi z} \tag{3.3}$$

and multiplication formula

$$\Gamma(z)\Gamma\left(z + \frac{1}{n}\right)\Gamma\left(z + \frac{2}{n}\right) \cdots \Gamma\left(z + \frac{n-1}{n}\right) = (2\pi)^{(n-1)/2} n^{-nz+1/2} \Gamma(nz), \tag{3.4}$$

follow straight from the defining product.

Exercise 3.1. Prove equations (3.2)–(3.4).

We also take for granted from a complex analysis course the evaluation

$$\int_0^\infty e^{-t}t^{z-1}\mathrm{d}t = \Gamma(z) \tag{3.5}$$

of the Eulerian integral (of the second kind) in the domain $\operatorname{Re} z > 0$.

Proposition 3.1. *The logarithmic derivative $\psi(z) = \Gamma'(z)/\Gamma(z)$ of the gamma function serves a generating function for the values of Riemann's zeta function at positive integers. More specifically,*

$$\psi(1-z) = -\gamma - \sum_{m=1}^{\infty} \zeta(m+1)z^m \quad \textit{for } |z| < 1.$$

Proof. It follows from the logarithmic differentiation of (3.1) that

$$-\psi(z) = \frac{1}{z} + \gamma + \sum_{k=1}^{\infty}\left(-\frac{1}{k} + \frac{1}{k(1+z/k)}\right)$$

for $z \neq 0, -1, -2, \ldots$. Furthermore, from (3.2) we have $\psi(1+z) = 1/z + \psi(z)$. Thus,

$$\begin{aligned} -\psi(1-z) = \frac{1}{z} - \psi(-z) &= \gamma + \sum_{k=1}^{\infty}\frac{1}{k}\left(-1 + \frac{1}{1-z/k}\right) \\ &= \gamma + \sum_{k=1}^{\infty}\frac{1}{k}\sum_{m=1}^{\infty}\left(\frac{z}{k}\right)^m = \gamma + \sum_{m=1}^{\infty} z^m \sum_{k=1}^{\infty}\frac{1}{k^{m+1}}, \end{aligned}$$

with all the internal series converging in the disk $|z| < 1$. □

Exercise 3.2. In this exercise we compute the Eulerian integral of the first kind

$$\mathrm{B}(\alpha,\beta) = \int_0^1 x^{\alpha-1}(1-x)^{\beta-1}\,\mathrm{d}x,$$

where $\operatorname{Re}\alpha > 0$ and $\operatorname{Re}\beta > 0$.

(a) Verify the following properties:

$$\mathrm{B}(\alpha,\beta) = \mathrm{B}(\beta,\alpha); \qquad \mathrm{B}(\alpha,\beta+1) = \frac{\beta}{\alpha}\,\mathrm{B}(\alpha+1,\beta);$$

$$\mathrm{B}(\alpha,\beta) = \mathrm{B}(\alpha+1,\beta) + \mathrm{B}(\alpha,\beta+1); \qquad \mathrm{B}(\alpha,\beta+1) = \frac{\beta}{\alpha+\beta}\,\mathrm{B}(\alpha,\beta).$$

(b) Show that

$$\Gamma(\alpha)\Gamma(\beta) = 4\lim_{R\to\infty}\iint_{[0,R]^2} f(x,y)\,\mathrm{d}x\,\mathrm{d}y = 4\lim_{R\to\infty}\iint_{S_R} f(x,y)\,\mathrm{d}x\,\mathrm{d}y$$

where $f(x,y) = e^{-(x^2+y^2)}x^{2\alpha-1}y^{2\beta-1}$ and S_R is the circular sector $x^2+y^2 \le R$, $x \ge 0$, $y \ge 0$.

(c) Pass to the polar coordinates $x = r\cos\theta$, $y = r\sin\theta$ in the integral

$$\iint_{S_R} f(x,y)\,\mathrm{d}x\,\mathrm{d}y$$

and use part (b) to conclude that

$$\mathrm{B}(\alpha,\beta) = \frac{\Gamma(\alpha)\Gamma(\beta)}{\Gamma(\alpha+\beta)}.$$

Hint. (b) Write

$$\Gamma(\alpha) = \int_0^\infty e^{-t}t^{\alpha-1}\,\mathrm{d}t = 2\int_0^\infty e^{-x^2}x^{2\alpha-1}\,\mathrm{d}x = 2\lim_{R\to\infty}\int_0^R e^{-x^2}x^{2\alpha-1}\,\mathrm{d}x$$

and, similarly, for $\Gamma(\beta)$; then show that

$$\left|\iint_{[0,R]^2} f(x,y)\,\mathrm{d}x\,\mathrm{d}y - \iint_{S_R} f(x,y)\,\mathrm{d}x\,\mathrm{d}y\right| \to 0 \quad \text{as } R\to\infty. \qquad \square$$

Exercise 3.3. (a) Show the integral expansion

$$\psi(z) = -\gamma + \int_0^1 \frac{1-t^{z-1}}{1-t}\,\mathrm{d}t$$

in the half-plane $\operatorname{Re} z > 0$.

(b) Prove that, for $n = 1, 2, 3, \dots$,

$$\psi(n) = -\gamma + \sum_{k=1}^{n-1}\frac{1}{k}.$$

3.2 Hurwitz's zeta function

In order to analyse the properties of Riemann's zeta function we turn our attention to its slightly more general version

$$\zeta(s,a) = \sum_{n=0}^\infty \frac{1}{(a+n)^s} \tag{3.6}$$

known as *Hurwitz's zeta function.* In this expression we treat a as a real constant from the interval $0 < a \le 1$ (though one can allow a to vary over the real line, and even over the complex plane); again, the series in (3.6) defines an analytic function of s in the region $\operatorname{Re} s > 1$. Observe that $\zeta(s,1) = \zeta(s)$.

Proposition 3.2. *For* $\operatorname{Re} s > 1$,

$$\zeta(s,a) = \frac{1}{\Gamma(s)} \int_0^\infty \frac{x^{s-1}e^{-ax}}{1-e^{-x}}\,\mathrm{d}x.$$

Proof. We start with the following consequence of (3.5):

$$(a+n)^{-s}\Gamma(s) = \int_0^\infty x^{s-1}e^{-(n+a)x}\,\mathrm{d}x.$$

Taking $\delta > 0$, we have in the domain $\sigma = \operatorname{Re} s \ge 1+\delta$,

$$\begin{aligned}
\Gamma(s)\zeta(s,a) &= \lim_{N\to\infty} \sum_{n=0}^{N} \int_0^\infty x^{s-1}e^{-(n+a)x}\,\mathrm{d}x \\
&= \lim_{N\to\infty} \left(\int_0^\infty \frac{x^{s-1}e^{-ax}}{1-e^{-x}}\,\mathrm{d}x - \int_0^\infty \frac{x^{s-1}e^{-(N+1+a)x}}{1-e^{-x}}\,\mathrm{d}x \right) \\
&= \lim_{N\to\infty} \left(\int_0^\infty \frac{x^{s-1}e^{-ax}}{1-e^{-x}}\,\mathrm{d}x - \int_0^\infty \frac{x^{s-1}e^{-(N+a)x}}{e^{x}-1}\,\mathrm{d}x \right).
\end{aligned}$$

Since $e^x \ge 1+x$ for $x \ge 0$, the absolute value of the second integral is estimated from above by the quantity

$$\int_0^\infty x^{\sigma-2}e^{-(N+a)x}\,\mathrm{d}x = (a+N)^{1-\sigma}\Gamma(\sigma-1),$$

which clearly tends to 0 as $N \to \infty$ in view of $\sigma - 1 \ge \delta > 0$. This gives the desired formula for $\operatorname{Re} s \ge 1+\delta$, hence for $\operatorname{Re} s > 1$. □

For real $\rho > 0$ (possibly, $\rho = \infty$), introduce a (Hankel-type) contour $D = D(\rho)$, which starts at $z = \rho$, passes once around the origin into the positive direction (without crossing the half-line $z \ge 0$) and ends up at $z = \rho$. Our principal interest is in the integral

$$\int_{D(\infty)} \frac{(-z)^{s-1}e^{-az}}{1-e^{-z}}\,\mathrm{d}z = \lim_{\rho\to\infty} \int_{D(\rho)} \frac{(-z)^{s-1}e^{-az}}{1-e^{-z}}\,\mathrm{d}z$$

for a fixed s from the half-plane $\sigma = \operatorname{Re} s \ge 1+\delta$. To avoid the unwanted poles of the integrand, we further assume that the contours $D(\rho)$ do not contain the points $\pm 2\pi i n$ for $n = 1, 2, \dots$. We specify the branch of $(-z)^{s-1} = e^{(s-1)\log(-z)}$ by choosing the $\log(-z)$ to be real for negative z; then $-\pi \le \arg(-z) \le \pi$ on the contours — this makes the integrand a single-valued function on $D(\rho)$. Of course, the integrand is not analytic inside $D(\rho)$ but we can still deform it within $\mathbb{C} \setminus [0,\infty)$ to the contour going along the upper bank of the cut $[0,\infty)$ from ρ to $\varepsilon > 0$, then making a circle of radius ε around the origin and finally returning from ε to ρ along the lower bank of the cut. At the beginning we have $\arg(-z) = -\pi$, so that $(-z)^{s-1} = e^{-\pi i(s-1)}z^{s-1}$, and at the end we get $\arg(-z) = \pi$, hence $(-z)^{s-1} = e^{\pi i(s-1)}z^{s-1}$. We set $-z = \varepsilon e^{i\theta}$ on the circle. Therefore,

$$\begin{aligned}
&\int_{D(\rho)} \frac{(-z)^{s-1}e^{-az}}{1-e^{-z}}\,\mathrm{d}z \\
&\quad = e^{-\pi i(s-1)}\int_\rho^\varepsilon \frac{x^{s-1}e^{-ax}}{1-e^{-x}}\,\mathrm{d}x + i\int_{-\pi}^{\pi} \frac{(\varepsilon e^{i\theta})^s e^{a\varepsilon(\cos\theta+i\sin\theta)}}{1-e^{\varepsilon(\cos\theta+i\sin\theta)}}\,\mathrm{d}\theta \\
&\qquad + e^{\pi i(s-1)}\int_\varepsilon^\rho \frac{x^{s-1}e^{-ax}}{1-e^{-x}}\,\mathrm{d}x \\
&\quad = -2i\sin\pi s\int_\varepsilon^\rho \frac{x^{s-1}e^{-ax}}{1-e^{-x}}\,\mathrm{d}x + i\varepsilon^{s-1}\int_{-\pi}^{\pi} \frac{\varepsilon e^{is\theta+a\varepsilon(\cos\theta+i\sin\theta)}}{1-e^{\varepsilon(\cos\theta+i\sin\theta)}}\,\mathrm{d}\theta
\end{aligned}$$

for $0 < \varepsilon \le \rho$. As $\varepsilon \to 0$ we have $\varepsilon^{s-1} \to 0$ and

$$\int_{-\pi}^{\pi} \frac{\varepsilon e^{is\theta+a\varepsilon(\cos\theta+i\sin\theta)}}{1-e^{\varepsilon(\cos\theta+i\sin\theta)}}\,\mathrm{d}\theta \to \int_{-\pi}^{\pi} \frac{e^{is\theta}}{\cos\theta+i\sin\theta}\,\mathrm{d}\theta = \int_{-\pi}^{\pi} e^{i(s-1)\theta}\,\mathrm{d}\theta,$$

since the integrand uniformly converges to its limit. We conclude that

$$\int_{D(\rho)} \frac{(-z)^{s-1}e^{-az}}{1-e^{-z}}\,\mathrm{d}z = -2i\sin\pi s\int_0^\rho \frac{x^{s-1}e^{-ax}}{1-e^{-x}}\,\mathrm{d}x$$

implying

$$\begin{aligned}
\int_{D(\infty)} \frac{(-z)^{s-1}e^{-az}}{1-e^{-z}}\,\mathrm{d}z &= -2i\sin\pi s\int_0^\infty \frac{x^{s-1}e^{-ax}}{1-e^{-x}}\,\mathrm{d}x \\
&= -2i\sin\pi s\,\Gamma(s)\zeta(s,a) = -2\pi i\,\frac{\zeta(s,a)}{\Gamma(1-s)}
\end{aligned}$$

on the basis of Proposition 3.2 and reflection formula (3.3). This brings us to the following result.

Proposition 3.3. *For* $\operatorname{Re} s > 1$,

$$\zeta(s,a) = -\frac{\Gamma(1-s)}{2\pi i}\int_{D(\infty)} \frac{(-z)^{s-1}e^{-az}}{1-e^{-z}}\,\mathrm{d}z. \tag{3.7}$$

The resulting integral is a single-valued analytic function of s for *all* $s \in \mathbb{C}$. Therefore, the only potential singularities of $\zeta(s,a)$ originate from the singularities of $\Gamma(1-s)$, which are the points $s = 1, 2, \dots$, since the integral provide the analytic continuation of $\zeta(s,a)$ to the entire complex plane with the exception of these points. At the same time, we already now the analyticity of $\zeta(s,a)$ in the domain $\operatorname{Re} s > 1$ from its defining series expansion (3.6). This leads us to the following.

Proposition 3.4. *The function $\zeta(s,a)$ is analytic in $\mathbb{C}$ besides $s=1$, where it has a simple pole with residue* 1.

When $a=1$, this implies the analytic properties of $\zeta(s)$.

Proof. By the argument above, the point $s=1$ is the only candidate for a singular point. Taking $s=1$ in the integral (without the gamma prefactor) we get the expression

$$\frac{1}{2\pi i}\int_{D(\infty)} \frac{e^{-az}}{1-e^{-z}}\,\mathrm{d}z$$

which is equal to the residue of the integrand at $z=0$: this is clearly equal to 1. Combined with (3.7) this implies

$$\lim_{s\to 1}\frac{\zeta(s,a)}{\Gamma(1-s)}=-1.$$

It remains to recall that $\Gamma(1-s)$ has a simple pole at $s=1$ with residue -1. $\square$

Exercise 3.4. Show for $\operatorname{Re} s>0$,

$$(1-2^{1-s})\zeta(s)=\sum_{n=1}^{\infty}\frac{(-1)^{n-1}}{n^s}=\frac{1}{\Gamma(s)}\int_0^{\infty}\frac{x^{s-1}}{e^x+1}\,\mathrm{d}x.$$

Exercise 3.5. Show for $\operatorname{Re} s>1$,

$$(2^s-1)\zeta(s)=\zeta\left(s,\frac{1}{2}\right)=\frac{2^s}{\Gamma(s)}\int_0^{\infty}\frac{x^{s-1}e^x}{e^{2x}-1}\,\mathrm{d}x.$$

Exercise 3.6. Show for all $s\neq 1$,

$$\zeta(s)=-\frac{2^{1-s}\Gamma(1-s)}{2\pi i\,(2^{1-s}-1)}\int_{D(\infty)}\frac{(-z)^{s-1}}{e^z+1}\,\mathrm{d}z,$$

where the contour $D(\infty)$ does not contain inside the points $\pm\pi i,\pm 3\pi i,\pm 5\pi i,\dots$.

Proposition 3.5 (Hurwitz). *For $0<a\le 1$ and $\sigma=\operatorname{Re} s<0$,*

$$\zeta(s,a)=\frac{2\Gamma(1-s)}{(2\pi)^{1-s}}\left(\sin\frac{\pi s}{2}\sum_{n=1}^{\infty}\frac{\cos 2\pi a n}{n^{1-s}}+\cos\frac{\pi s}{2}\sum_{n=1}^{\infty}\frac{\sin 2\pi a n}{n^{1-s}}\right). \tag{3.8}$$

Proof. Consider the integral

$$-\frac{1}{2\pi i}\int_{C_N}\frac{(-z)^{s-1}e^{-az}}{1-e^{-z}}\,\mathrm{d}z,$$

where N is an *odd* positive integer, the contour C_N is the circle centred at the origin of radius $N\pi$ going counter-clockwise from $N\pi$ to $N\pi$. We assume that $\arg(-z) = 0$ at $z = -N\pi$.

In the domain bounded by the contours C_N and $D(N\pi)$, the function $(-z)^{s-1}e^{-az}/(1-e^{-z})$ is analytic and single-valued, except for the poles at $\pm 2\pi i, \pm 4\pi i, \ldots, \pm(N-1)\pi i$. Therefore,

$$\frac{1}{2\pi i}\int_{C_N}\frac{(-z)^{s-1}e^{-az}}{1-e^{-z}}\,\mathrm{d}z - \frac{1}{2\pi i}\int_{D(N\pi)}\frac{(-z)^{s-1}e^{-az}}{1-e^{-z}}\,\mathrm{d}z$$
$$= \sum_{n=1}^{(N-1)/2}(R_n^+ + R_n^-),$$

where R_n^+ and R_n^- are the residues of the integrand at $2n\pi i$ and $-2n\pi i$, respectively. When $-z = 2n\pi e^{-\pi i/2}$, the residue is equal to $(2n\pi)^{s-1}e^{-\pi i(s-1)/2}e^{-2an\pi i}$, so that

$$R_n^+ + R_n^- = 2\,(2n\pi)^{s-1}\sin\left(\frac{\pi s}{2} + 2\pi a n\right) \quad \text{for } n = 1, 2, \ldots, \frac{N-1}{2}.$$

We obtain

$$-\frac{1}{2\pi i}\int_{D(N\pi)}\frac{(-z)^{s-1}e^{-az}}{1-e^{-z}}\,\mathrm{d}z = \frac{2\sin\frac{\pi s}{2}}{(2\pi)^{1-s}}\sum_{n=1}^{(N-1)/2}\frac{\cos 2\pi a n}{n^{1-s}}$$
$$+ \frac{2\cos\frac{\pi s}{2}}{(2\pi)^{1-s}}\sum_{n=1}^{(N-1)/2}\frac{\sin 2\pi a n}{n^{1-s}} - \frac{1}{2\pi i}\int_{C_N}\frac{(-z)^{s-1}e^{-az}}{1-e^{-z}}\,\mathrm{d}z.$$

Furthermore, for $0 < a \leq 1$ we can find an absolute bound $|e^{-az}/(1-e^{-z})| < M$ for $z \in C_N$, independent of N. This means that, for $\sigma = \operatorname{Re} s < 0$,

$$\left|\frac{1}{2\pi i}\int_{C_N}\frac{(-z)^{s-1}e^{-az}}{1-e^{-z}}\,\mathrm{d}z\right| < \frac{M}{2\pi}\int_{-\pi}^{\pi}|(N\pi)^s e^{is\theta}|\,\mathrm{d}\theta$$
$$< M(N\pi)^\sigma e^{\pi|s|} \to 0 \quad \text{as } N \to \infty.$$

Thus, letting $N \to \infty$ in the above equality we arrive at the desired formula (3.8). Note the (absolute) convergence of the both series when $\operatorname{Re} s < 0$. □

Theorem 3.1 (Riemann). *The following functional equation is valid for Riemann's zeta function:*

$$2^{1-s}\Gamma(s)\zeta(s)\cos\frac{\pi s}{2} = \pi^s\zeta(1-s). \tag{3.9}$$

Proof. Take $a = 1$ in equation (3.8) and apply the reflection formula (3.3) of the gamma function. This proves (3.9) in the domain $\operatorname{Re} s < 0$. Since the both sides are analytic in the larger domain $\mathbb{C} \setminus \{0, 1\}$ (besides the simple poles at $s = 0, 1$), the result remains valid there by the theory of analytic continuation. □

Exercise 3.7. Show the function $\Gamma(s/2)\pi^{-s/2}\zeta(s)$ does not change under the involution $s \leftrightarrow 1 - s$.

It follows from (3.9) that $\zeta(s)$ has zeros at negative even integers; these are called trivial zeros. In his famous 1859 memoir, Riemann suggested that all other (non-trivial) zeros lie on the critical line $\operatorname{Re} s = 1/2$, which represents the symmetry of the functional equation.

3.3 Zeta values and Bernoulli numbers

One of interesting and still unsolved problems is the problem of determining polynomial relations over $\mathbb{Q}$ for the numbers $\zeta(s)$, $s = 2, 3, 4, \dots$.

The first breakthrough in this direction is due to Euler, who showed that $\zeta(2k)$ is always a rational multiple of π^{2k}, where

$$\begin{aligned}\pi &= 4\sum_{n=0}^{\infty} \frac{(-1)^n}{2n+1} \\ &= 3.14159265358979323846264338327950288419716939937510\dots.\end{aligned}$$

Although we do not follow Euler's original method, the derivation is worth reproducing.

For $a \in \mathbb{R}$, the *Bernoulli polynomials* $B_s(a) \in \mathbb{Q}[a]$, where $s = 0, 1, 2, \dots$, are defined by the generating function

$$\frac{ze^{az}}{e^z - 1} = \sum_{s=0}^{\infty} B_s(a)\,\frac{z^s}{s!}, \tag{3.10}$$

while the *Bernoulli numbers* $B_s \in \mathbb{Q}$, where $s = 0, 1, 2, \dots$, are simply given by $B_s = B_s(0)$. The latter means that the generating function of the Bernoulli numbers is

$$\frac{z}{e^z - 1} = \sum_{s=0}^{\infty} B_s\,\frac{z^s}{s!}.$$

For example, $B_0 = 1$, $B_1 = -1/2$. The polynomials and numbers satisfy numerous identities. As an example, we have the formulas $B_s'(a) = sB_{s-1}(a)$

and

$$\sum_{k=M}^{N-1} k^{s-1} = \frac{B_s(N) - B_s(M)}{s}$$

for $s = 1, 2, \ldots$, and also the following ones.

Exercise 3.8. (a) Show that

$$B_s(a) = \sum_{k=0}^{s} \binom{s}{k} B_k a^{s-k} \quad \text{for } s = 0, 1, 2, \ldots .$$

(b) Verify that $B_s = 0$ for odd $s \geq 3$.
(c) Verify that $B_s(1) = B_s = B_s(0)$ for even $s \geq 0$.

Lemma 3.1. *For* $0 < a \leq 1$ *and* $s = -m$ *a negative integer,*

$$\zeta(-m, a) = -\frac{B_{m+1}(a)}{m+1}.$$

Proof. Recall the integral

$$\frac{1}{2\pi i} \int_{D(\infty)} \frac{(-z)^{s-1} e^{-az}}{1 - e^{-z}} \, \mathrm{d}z = -\frac{\zeta(s, a)}{\Gamma(1 - s)}$$

from Proposition 3.3. If s is a negative integer, $s = -m$, the expression

$$\frac{(-z)^{s-1} e^{-az}}{1 - e^{-z}}$$

is a single-valued function of z, which is analytic in $|z| < 2\pi$, $z \neq 0$. By Cauchy's integral theorem, the integral over $D(\infty)$ is equal to the residue of the integrand at $z = 0$, that is, to the coefficient of $z^{-s} = z^m$ in

$$\frac{(-1)^{s-1} e^{-az}}{1 - e^{-z}} = \frac{(-1)^{s-1}}{z} \frac{(-z)\, e^{-az}}{e^{-z} - 1} = \frac{(-1)^{m-1}}{z} \sum_{k=0}^{\infty} (-1)^k B_k(a) \frac{z^k}{k!}.$$

It follows that

$$-\frac{\zeta(-m, a)}{m!} = -\frac{\zeta(s, a)}{\Gamma(1 - s)}\bigg|_{s=-m} = \frac{B_{m+1}(a)}{(m+1)!},$$

which implies the result. □

When $a = 1$, we get the following consequence for Riemann's zeta function (using also Exercise 3.8).

Proposition 3.6. *For* $k = 1, 2, \ldots$, *we have* $\zeta(-2k) = 0$ *and* $\zeta(1 - 2k) = B_{2k}/(2k)$.

Exercise 3.9. Show that $\zeta(0, a) = \frac{1}{2} - a$ and $\zeta(0) = -\frac{1}{2}$.

Proposition 3.7. *For $k = 1, 2, \dots$, we have*

$$\zeta(2k) = (-1)^{k-1} \frac{(2\pi)^{2k} B_{2k}}{2\,(2k)!}.$$

Proof. This follows from Proposition 3.6 and the functional equation (3.9) for $s = 2k$. □

In particular,

$$\zeta(2) = \frac{\pi^2}{2 \cdot 3}, \quad \zeta(4) = \frac{\pi^4}{2 \cdot 3^2 \cdot 5}, \quad \zeta(6) = \frac{\pi^6}{3^3 \cdot 5 \cdot 7},$$
$$\zeta(8) = \frac{\pi^8}{2 \cdot 3^3 \cdot 5^2 \cdot 7}, \quad \zeta(10) = \frac{\pi^{10}}{3^5 \cdot 5 \cdot 7 \cdot 11},$$
$$\zeta(12) = \frac{691\pi^{12}}{3^6 \cdot 5^3 \cdot 7^2 \cdot 11 \cdot 13}, \quad \zeta(14) = \frac{2\pi^{14}}{3^6 \cdot 5^2 \cdot 7 \cdot 11 \cdot 13},$$

and so on.

Proposition 3.7 gives us a 'closed form' expression for the values of the zeta function at even integers in terms of π and the (rational) Bernoulli numbers. No similar formulae are known for the values at odd integers. In Chapter 7 we touch questions about arithmetic nature of *zeta values* — the values of $\zeta(s)$ at integers $s \ge 2$; see there Conjecture 7.1.

The difficulty of proving that the 'odd' zeta values $\zeta(3), \zeta(5), \zeta(7), \dots$ are algebraically independent with π over $\mathbb{Q}$ serves a motivation to introducing a multidimensional generalization of Riemann's zeta function. For positive integers $s_1, s_2, \dots, s_l$ with $s_1 > 1$, consider the values of the multiple (l-tuple) zeta function

$$\zeta(\boldsymbol{s}) = \zeta(s_1, s_2, \dots, s_l) = \sum_{n_1 > n_2 > \cdots > n_l \ge 1} \frac{1}{n_1^{s_1} n_2^{s_2} \cdots n_l^{s_l}}; \tag{3.11}$$

the corresponding multi-index $\boldsymbol{s} = (s_1, s_2, \dots, s_l)$ will be further regarded as *admissible*. The quantities (3.11) are called the *multiple zeta values* (and abbreviated MZVs), or the *multiple harmonic series*, or the *Euler sums*. The sums (3.11) for $l = 2$ were first investigated by Euler, who obtained a family of identities connecting double and ordinary zeta values. In particular, Euler proved the identity

$$\zeta(2, 1) = \zeta(3), \tag{3.12}$$

which was several times rediscovered by others later.

Exercise 3.10. Find your own (elementary) proof of (3.12).

The following exercise discusses q-deformations of (multiple) zeta values (see also Section 1.3).

Exercise 3.11. Let $\sigma_k(n) = \sum_{d|n} d^k$ the sum of the kth powers of the divisors of n. In this exercise we assume that q is a complex parameter from the unit disk $|q| < 1$.

(a) Show that $\sigma_k(n)$ is a multiplicative function (see Section 2.4 and compare with Exercise 1.1).

(b) Prove that

$$\sum_{n=1}^{\infty} \sigma_k(n) q^n = \sum_{n=1}^{\infty} \frac{n^k q^n}{1-q^n}.$$

(c) Prove that

$$\sum_{n=1}^{\infty} \sigma_1(n) q^n = \sum_{n=1}^{\infty} \frac{q^n}{(1-q^n)^2}$$

and deduce from this that

$$\lim_{q\to 1} (1-q)^2 \sum_{n=1}^{\infty} \sigma_1(n) q^n = \zeta(2).$$

(d) Prove that

$$\sum_{n=1}^{\infty} \sigma_2(n) q^n = \sum_{n=1}^{\infty} \frac{q^n(1+q^n)}{(1-q^n)^3}$$

and deduce from this that

$$\lim_{q\to 1} (1-q)^3 \sum_{n=1}^{\infty} \sigma_2(n) q^n = 2\zeta(3).$$

(e) Demonstrate that

$$\sum_{n=1}^{\infty} \frac{q^n}{(1-q^n)^2} \sum_{\ell=1}^{n} \frac{1}{1-q^\ell} = \sum_{n=1}^{\infty} \sigma_2(n) q^n$$

and that the limiting case as $q \to 1$ of this identity after both sides are multiplied by $(1-q)^3$ is precisely Euler's relation (3.12).

(f) Generalise identities from parts (c) and (d) to the form

$$\sum_{n=1}^{\infty} \sigma_k(n) q^n = \sum_{n=1}^{\infty} \frac{P_k(q^n)}{(1-q^n)^{k+1}}$$

and compute the limit

$$\lim_{q\to 1} (1-q)^{k+1} \sum_{n=1}^{\infty} \sigma_k(n) q^n.$$

3.4 Analytic continuation of multiple zeta function

In this part, we discuss analytic properties of the *multiple zeta function* (MZF)

$$\zeta(\boldsymbol{s}) = \sum_{n_1>n_2>\cdots>n_l\ge 1} \frac{1}{n_1^{s_1} n_2^{s_2} \cdots n_l^{s_l}} \tag{3.13}$$

as a function of complex variables $s_1, \dots, s_l$; the notation $\sigma_1, \dots, \sigma_l$ will be used for the real parts of $s_1, \dots, s_l$.

Exercise 3.12. Show that the multiple series in (3.13) converges *absolutely* in the domain

$$\sigma_1 + \cdots + \sigma_j = \mathrm{Re}(s_1 + \cdots + s_j) > j \quad \text{for every } j = 1, \dots, l.$$

Conclude from this that the MZV is analytic in each of its variables in the domain $\sigma_1 + \cdots + \sigma_j > j$, where $j = 1, \dots, l$.

Hint. Use mathematical induction on l and estimates

$$\sum_{n>M} \frac{1}{n^\sigma} \le \frac{1}{(\sigma-1)M^{\sigma-1}},$$

where $M \ge 1$ is integral and $\sigma > 1$ is real, coming from the integral test (when the partial sums of a series are compared to Riemann sums). □

Lemma 3.2. *For $0 < a \le 1$ and an integer $m \ge 2$,*

$$\sum_{n\in\mathbb{Z}}{}' \frac{e^{2\pi i n a}}{(2\pi i n)^m} = -\frac{B_m(a)}{m!},$$

where the dash in summation corresponds to omitting the (problematic) index $n = 0$.

Proof. Comparing Hurwitz's equation (3.8),

$$\frac{\zeta(s,a)}{\Gamma(1-s)} = \frac{2}{(2\pi)^{1-s}} \sum_{n=1}^{\infty} \frac{\sin(\pi s/2 + 2\pi a n)}{n^{1-s}}$$

for $s = -m+1$, with the result of Lemma 3.1,

$$-\frac{B_m(a)}{m!} = \frac{\zeta(s,a)}{\Gamma(1-s)}\bigg|_{s=-m+1},$$

we find

$$-\frac{B_m(a)}{m!} = 2\sum_{n=1}^{\infty} \frac{\sin(-\pi(m-1)/2 + 2\pi a n)}{(2\pi n)^m},$$

which is exactly

$$(-1)^k \sum_{n=1}^{\infty} \frac{2\sin 2\pi an}{(2\pi n)^{2k+1}} = \sum_{n=1}^{\infty} \frac{e^{2\pi ina} - e^{-2\pi ina}}{(2\pi in)^{2k+1}}$$
$$= \sum_{n=1}^{\infty} \frac{e^{2\pi ina}}{(2\pi in)^{2k+1}} + \sum_{n=1}^{\infty} \frac{e^{-2\pi ina}}{(-2\pi in)^{2k+1}}$$

or

$$(-1)^k \sum_{n=1}^{\infty} \frac{2\cos 2\pi an}{(2\pi n)^{2k}} = \sum_{n=1}^{\infty} \frac{e^{2\pi ina} + e^{-2\pi ina}}{(2\pi in)^{2k}}$$
$$= \sum_{n=1}^{\infty} \frac{e^{2\pi ina}}{(2\pi in)^{2k}} + \sum_{n=1}^{\infty} \frac{e^{-2\pi ina}}{(-2\pi in)^{2k}}$$

depending on whether $m = 2k+1$ is odd or $m = 2k$ is even. □

Lemma 3.3. *For $0 < a \le 1$ and any integer $m \ge 2$,*

$$|B_m(a)| < \frac{4m!}{(2\pi)^m}.$$

Proof. It follows from Lemma 3.3 that

$$|B_m(a)| \le m! \sum_{n\in\mathbb{Z}}{}' \frac{1}{(2\pi n)^m} = \frac{2m!\,\zeta(m)}{(2\pi)^m}.$$

It remains to apply the trivial estimate $\zeta(m) \le \zeta(2) = \pi^2/6 < 2$. □

For the statement and application of the following classical result, it will be convenient to introduce the *periodic* Bernoulli polynomials given by $\widetilde{B}_m(a) = B_m(\{a\})$, where $\{\cdot\}$ denotes the fractional part of a real number. By Lemma 3.3 (and Exercise 3.8) we get the estimate

$$|\widetilde{B}_m(a)| < \frac{4m!}{(2\pi)^m} \quad \text{for } m = 2, 3, \dots, \tag{3.14}$$

now valid for *all* real a.

Exercise 3.13. Verify the validity of (3.14) for $m = 0, 1$.

We will also implement the (standard) notation

$$(s)_m = \frac{\Gamma(s+m)}{\Gamma(s)} = \begin{cases} s(s+1)\cdots(s+m-1) & \text{if } m = 1, 2, \dots, \\ 1 & \text{if } m = 0, \end{cases} \tag{3.15}$$

for the Pochhammer symbol, though it makes sense for *any* (not necessarily integer or non-negative) m. For example, $(s)_{-1} = \Gamma(s-1)/\Gamma(s) = 1/(s-1)$.

Proposition 3.8 (Euler–Maclaurin summation). *Let $f(x)$ be a (complex-valued) C^∞ function on the real interval $[1,\infty)$. Then for any positive integers N and m, m even,*

$$\sum_{n=1}^{N} f(n) = \int_1^N f(x)\,\mathrm{d}x + \frac{1}{2}\big(f(1)+f(N)\big) + \sum_{k=2}^{m} \frac{B_k}{k!}\left(f^{(k-1)}(N) - f^{(k-1)}(1)\right)$$
$$- \frac{1}{m!}\int_1^N \widetilde{B}_m(x) f^{(m)}(x)\,\mathrm{d}x.$$

Notice that the sum over k in the formula only involves k even, because $B_k = 0$ for odd $k \geq 2$.

Lemma 3.4. *Given $s \in \mathbb{C}$ with $\operatorname{Re} s > 1$, for integers $M \geq 1$ and $m \geq 2$, m even, we have*

$$\sum_{n>M} \frac{1}{n^s} = \sum_{k=0}^{m} \frac{B_k}{k!}\frac{(s)_{k-1}}{M^{s+k-1}} - \frac{(s)_m}{m!}\int_M^\infty \frac{\widetilde{B}_m(x)}{x^{s+m}}\,\mathrm{d}x.$$

Proof. Apply Proposition 3.8 with $f(x) = 1/x^s$ twice: when $N \to \infty$ and when $N = M$. Taking the difference of the results we arrive at

$$\begin{aligned}\sum_{n>M}\frac{1}{n^s} &= \sum_{n=1}^{\infty}\frac{1}{n^s} - \sum_{n=1}^{M}\frac{1}{n^s}\\ &= \int_M^\infty f(x)\,\mathrm{d}x - \frac{1}{2}f(M) - \sum_{k=2}^{m}\frac{B_k}{k!}f^{(k-1)}(M)\\ &\quad - \frac{1}{m!}\int_M^\infty \widetilde{B}_m(x) f^{(m)}(x)\,\mathrm{d}x\\ &= \frac{1}{(s-1)M^{s-1}} - \frac{1}{2M^s} - \sum_{k=2}^{m}\frac{B_k}{k!}\frac{(s)_{k-1}}{M^{s+k-1}} - \frac{(s)_m}{m!}\int_M^\infty \frac{\widetilde{B}_m(x)}{x^{s+m}}\,\mathrm{d}x,\end{aligned}$$

which can be written in the desired form because $B_0 = 1$ and $B_1 = -1/2$. □

Exercise 3.14. Use Lemma 3.4 (with $M = 1$, say) and the estimates of Lemma 3.3 to show that Riemann's zeta function can be analytically continued to the half-plane $\operatorname{Re} s > -L$ for any real $L > 0$.

Introduce the following discrete subset of $\mathbb{C}^l$:

$$\Sigma_l = \big\{ \boldsymbol{s} \in \mathbb{C}^l : s_1 \in \{1\},\ s_1 + s_2 \in \{1,2\} \cup 2\mathbb{Z}_{\leq 0},$$
$$s_1 + \cdots + s_j \in \mathbb{Z}_{\leq j} \text{ for } j = 3,\ldots,l \big\}.$$

The following general result provides the analytic continuation of the MZV $\zeta(\boldsymbol{s})$ to a meromorphic function on $\mathbb{C}^l$ with (at most) simple poles given by Σ_l.

Theorem 3.2. *Assume $l \geq 2$. Then for any $\boldsymbol{s} = (s_1, \dots, s_l) \in \mathbb{C}^l \setminus \Sigma_l$ and an even $m > l + |\sigma_1| + \cdots + |\sigma_l|$, we have*

$$\zeta(\boldsymbol{s}) = \sum_{k=0}^{m} \frac{B_k}{k!}(s_1)_{k-1} \cdot \zeta(s_1 + s_2 + k - 1, s_3, \dots, s_l) \\ - \frac{(s_1)_m}{m!} \sum_{n_2 > \cdots > n_l \geq 1} \frac{1}{n_2^{s_2} \cdots n_l^{s_l}} \int_{n_2}^{\infty} \frac{\widetilde{B}_m(x)}{x^{s_1+m}} \, \mathrm{d}x. \tag{3.16}$$

Proof. The absolute convergence of the second series in the formula (3.16) follows from the estimate

$$\left| \int_M^{\infty} \frac{\widetilde{B}_m(x)}{x^{s+m}} \, \mathrm{d}x \right| \leq \frac{4m!}{(2\pi)^{2m}} \int_M^{\infty} \frac{\mathrm{d}x}{x^{\sigma+m}} = \frac{4m!}{(2\pi)^{2m}(m-1+\sigma)M^{m-1+\sigma}},$$

where $\sigma = \operatorname{Re} s$, implying

$$\sum_{n_2 > \cdots > n_l \geq 1} \left| \frac{1}{n_2^{s_2} n_3^{s_3} \cdots n_l^{s_l}} \int_{n_2}^{\infty} \frac{\widetilde{B}_m(x)}{x^{s_1+m}} \, \mathrm{d}x \right| \\ \leq \frac{4m!}{(2\pi)^{2m}(m-1+\sigma_1)} \sum_{n_2 > \cdots > n_l \geq 1} \frac{1}{n_2^{m-1+\sigma_1+\sigma_2} n_3^{\sigma_3} \cdots n_l^{\sigma_l}}.$$

For the latter sum we use

$$\frac{1}{n_2^{\sigma_1+\sigma_2} n_3^{\sigma_3} \cdots n_l^{\sigma_l}} \leq n_2^{|\sigma_1|+|\sigma_2|} n_3^{|\sigma_3|} \cdots n_l^{|\sigma_l|} \leq n_2^{|\sigma_1|+|\sigma_2|+|\sigma_3|+\cdots+|\sigma_l|}$$

and the fact that the number of integers $n_3, \dots, n_l$ satisfying $n_2 > n_3 > \cdots > n_l \geq 1$ is bounded above by n_2^{l-2} (because each n_j satisfies $1 \leq n_j < n_2$), so that

$$\sum_{n_2 > \cdots > n_l \geq 1} \frac{1}{n_2^{m-1+\sigma_1+\sigma_2} n_3^{\sigma_3} \cdots n_l^{\sigma_l}} \leq \sum_{n_2 \geq 1} \frac{n_2^{|\sigma_1|+|\sigma_2|+|\sigma_3|+\cdots+|\sigma_l|} n_2^{l-2}}{n_2^{m-1}}$$

converges when $m > l + |\sigma_1| + \cdots + |\sigma_l|$.

Now, to get the formula (3.16) we apply Lemma 3.4 with $s = s_1$, $n = n_1$ and $M = n_2$, and then perform the summation over $n_2 > n_3 > \cdots > n_l \geq 1$.

It remains to carefully control the (potential) poles by induction on l. □

Exercise 3.15. Show that the potential poles of $\zeta(\boldsymbol{s})$ at $\boldsymbol{s} \in \Sigma_l$ are at most simple.

Hint. Notice that the second (multiple) sum in (3.16) is analytic, so that the only source for poles comes from

$$\sum_{k=0}^{m} \frac{B_k}{k!} (s_1)_{k-1} \cdot \zeta(s_1 + s_2 + k - 1, s_3, \ldots, s_l).$$

Use mathematical induction on l and the fact that $\zeta(s)$ (when $l = 1$) has one simple pole at $s = 1$. □

Chapter notes

Though the Whittaker–Watson textbook [80] remains our principal recommendation for treatment of basic special functions, the book [77] is a minimalistic alternative, especially if planned as a real-life university course. When it comes to the Bernoulli numbers and polynomials and their connection with zeta functions of various types, there is a way more to say — the book [5] is a tremendous source on this topic.

The multiple zeta values and their generalisations have received a very special attention and are under extensive studies during the last decades, in connection with problems of not only number theory but also of combinatorics, algebra, analysis, algebraic geometry, quantum physics, and many other branches of mathematics. This is a topic of its own, with many results and challenges left; the interested reader is advised to consult with the books [20, 87].

Chapter 4

Continued fractions

4.1 Euclidean algorithm and continuants

Let $a, b \in \mathbb{Z}$ with $a > b > 0$. Defining $r_1 = b$, consider the following successive application of division with remainder:

$$\begin{aligned} a &= a_1 r_1 + r_2, & 0 < r_2 < r_1, \\ r_1 &= a_2 r_2 + r_3, & 0 < r_3 < r_2, \\ &\vdots \\ r_{n-2} &= a_{n-1} r_{n-1} + r_n, & 0 < r_n < r_{n-1}, \\ r_{n-1} &= a_n r_n + 0. \end{aligned} \tag{4.1}$$

Then the last non-zero remainder r_n is the greatest common divisor of a and b.

Critically, the procedure (4.1) terminates at some step in view of the following chain of inequalities:

$$b = r_1 > r_2 > \cdots > r_{n-1} > r_n > 0.$$

Also observe that on the last step we get $a_n \geq 2$, as otherwise ($a_n = 1$) we would have $r_{n-1} = r_n$ contradicting $r_{n-1} < r_n$.

By applying consequently the steps of the Euclidean algorithm we deduce the representation

$$\frac{a}{b} = a_1 + \frac{r_2}{r_1} = a_1 + \frac{1}{r_1/r_2} = \cdots = a_1 + \cfrac{1}{a_2 + \cfrac{1}{a_3 + \ddots + \cfrac{1}{a_n}}}, \tag{4.2}$$

where $a_1, \ldots, a_n \in \mathbb{Z}_{>0}$ (with $a_n \geq 2$) are the *partial quotients* of the *finite continued fraction* for a/b. Notice that the intermediate divisors

$r_2, \ldots, r_n$ do not appear in the continued fraction representation, so that we can restrict ourselves to the case of a and b relatively prime, $(a, b) = 1$. Furthermore, the representation (4.2) with $a_n \geq 2$ is unique for $a/b > 1$ rational, because it is in a one-to-one correspondence with the steps in the Euclidean algorithm.

Consider now $a_1, a_2, a_3, \ldots$ as unknowns (variables) and define polynomials $p_n = p_n(a_1, \ldots, a_n)$ and $q_n = q_n(a_1, \ldots, a_n)$ as follows: $p_1(a_1) = a_1$, $q_1(a_1) = 1$ and

$$\begin{aligned} p_n(a_1, a_2, \ldots, a_n) &= a_1 p_{n-1}(a_2, \ldots, a_n) + q_{n-1}(a_2, \ldots, a_n), \\ q_n(a_1, a_2, \ldots, a_n) &= p_{n-1}(a_2, \ldots, a_n) \end{aligned} \tag{4.3}$$

for $n = 2, 3, \ldots$. One can also start (4.3) from $n = 1$ by setting $p_0(\,) = 1$ and $q_0(\,) = 0$.

Lemma 4.1. *For* $n = 1, 2, \ldots,$ *we have*

$$\frac{p_n(a_1, a_2, \ldots, a_n)}{q_n(a_1, a_2, \ldots, a_n)} = a_1 + \cfrac{1}{a_2 + \cfrac{1}{a_3 + \ddots + \cfrac{1}{a_n}}}.$$

Proof. This follows by induction on n from noticing that $p_1(a_1)/q_1(a_1) = a_1$ and

$$\begin{aligned} \frac{p_n(a_1, a_2, \ldots, a_n)}{q_n(a_1, a_2, \ldots, a_n)} &= \frac{a_1 p_{n-1}(a_2, \ldots, a_n) + q_{n-1}(a_2, \ldots, a_n)}{p_{n-1}(a_2, \ldots, a_n)} \\ &= a_1 + \frac{1}{p_{n-1}(a_2, \ldots, a_n)/q_{n-1}(a_2, \ldots, a_n)}. \end{aligned}$$ □

Lemma 4.2. *We have the matrix identity*

$$\begin{pmatrix} p_n(a_1, a_2, \ldots, a_n) \\ q_n(a_1, a_2, \ldots, a_n) \end{pmatrix} = \begin{pmatrix} a_1 & 1 \\ 1 & 0 \end{pmatrix} \begin{pmatrix} p_{n-1}(a_2, \ldots, a_n) \\ q_{n-1}(a_2, \ldots, a_n) \end{pmatrix}.$$

Proof. This is just another way to write (4.3). □

Iterating the identity of Lemma 4.2 we obtain

$$\begin{aligned} \begin{pmatrix} p_n(a_1, a_2, \ldots, a_n) \\ q_n(a_1, a_2, \ldots, a_n) \end{pmatrix} &= \begin{pmatrix} a_1 & 1 \\ 1 & 0 \end{pmatrix} \begin{pmatrix} a_2 & 1 \\ 1 & 0 \end{pmatrix} \cdots \begin{pmatrix} a_{n-1} & 1 \\ 1 & 0 \end{pmatrix} \begin{pmatrix} a_n \\ 1 \end{pmatrix} \\ &= \begin{pmatrix} a_1 & 1 \\ 1 & 0 \end{pmatrix} \begin{pmatrix} a_2 & 1 \\ 1 & 0 \end{pmatrix} \cdots \begin{pmatrix} a_{n-1} & 1 \\ 1 & 0 \end{pmatrix} \begin{pmatrix} a_n & 1 \\ 1 & 0 \end{pmatrix} \begin{pmatrix} 1 \\ 0 \end{pmatrix}. \end{aligned}$$

Using this we arrive at

Lemma 4.3 (key identity). *For $p_n = p_n(a_1, \ldots, a_n)$, $q_n = q_n(a_1, \ldots, a_n)$, $p_{n-1} = p_{n-1}(a_1, \ldots, a_{n-1})$ and $q_{n-1} = q_{n-1}(a_1, \ldots, a_{n-1})$, we have the matrix identity*

$$\begin{pmatrix} p_n & p_{n-1} \\ q_n & q_{n-1} \end{pmatrix} = \begin{pmatrix} a_1 & 1 \\ 1 & 0 \end{pmatrix} \begin{pmatrix} a_2 & 1 \\ 1 & 0 \end{pmatrix} \cdots \begin{pmatrix} a_{n-1} & 1 \\ 1 & 0 \end{pmatrix} \begin{pmatrix} a_n & 1 \\ 1 & 0 \end{pmatrix}.$$

In particular, this implies

$$\begin{pmatrix} p_n & p_{n-1} \\ q_n & q_{n-1} \end{pmatrix} = \begin{pmatrix} p_{n-1} & p_{n-2} \\ q_{n-1} & q_{n-2} \end{pmatrix} \begin{pmatrix} a_n & 1 \\ 1 & 0 \end{pmatrix},$$

hence for the first column on the left-hand side

$$p_n = a_n p_{n-1} + p_{n-2},$$
$$q_n = a_n q_{n-1} + q_{n-2},$$

for $n = 2, 3, 4, \ldots$, where all the polynomials involved depend on $a_1, a_2, \ldots$ (without the shift!).

Finally, we call the polynomial

$$C(a_1, a_2, \ldots, a_n) = p_n(a_1, a_2, \ldots, a_n)$$

the *continuant* on variables $a_1, \ldots, a_n$. (We also set $C(\,) = 1$ for the empty set of variables.) It follows from (4.3) that

$$q_n(a_1, \ldots, a_n) = C(a_2, \ldots, a_n),$$

so that Lemma 4.1 reads

$$\frac{C(a_1, a_2, \ldots, a_n)}{C(a_2, \ldots, a_n)} = a_1 + \cfrac{1}{a_2 + \cfrac{1}{a_3 + \ddots + \cfrac{1}{a_n}}}.$$

Furthermore, the key identity assumes the form

$$\begin{pmatrix} C(a_1, a_2, \ldots, a_{n-1}, a_n) & C(a_1, a_2, \ldots, a_{n-1}) \\ C(a_2, \ldots, a_{n-1}, a_n) & C(a_2, \ldots, a_{n-1}) \end{pmatrix}$$
$$= \begin{pmatrix} a_1 & 1 \\ 1 & 0 \end{pmatrix} \begin{pmatrix} a_2 & 1 \\ 1 & 0 \end{pmatrix} \cdots \begin{pmatrix} a_{n-1} & 1 \\ 1 & 0 \end{pmatrix} \begin{pmatrix} a_n & 1 \\ 1 & 0 \end{pmatrix}, \qquad (4.4)$$

and we have the followings properties of the continuant.

Lemma 4.4. *For $n = 2, 3, \ldots$ and $a_i \in \mathbb{Z}_{>0}$, we have*

(a) $C(a_1, a_2, \ldots, a_n) = C(a_n, \ldots, a_2, a_1)$;
(b) $C(a_1, \ldots, a_s, a_{s+1}, \ldots, a_n) = C(a_1, \ldots, a_s)\, C(a_{s+1}, \ldots, a_n) + C(a_1, \ldots, a_{s-1})\, C(a_{s+2}, \ldots, a_n)$.

We can refer to property (a) as *reflection*, and to (b) as *reduction*.

Proof. For (a), use the matrix transposition of the key identity (4.4). For (b), write the identity in the form

$$\begin{pmatrix} C(a_1,\dots,a_n) & C(a_1,\dots,a_{n-1}) \\ C(a_2,\dots,a_n) & C(a_2,\dots,a_{n-1}) \end{pmatrix} = \begin{pmatrix} C(a_1,\dots,a_s) & C(a_1,\dots,a_{s-1}) \\ C(a_2,\dots,a_s) & C(a_2,\dots,a_{s-1}) \end{pmatrix}$$
$$\times \begin{pmatrix} C(a_{s+1},\dots,a_n) & C(a_{s+1},\dots,a_{n-1}) \\ C(a_{s+2},\dots,a_n) & C(a_{s+2},\dots,a_{n-1}) \end{pmatrix},$$

and read off the $(1,1)$-coefficient of the matrix on the left- and right-hand sides of the result. □

Exercise 4.1 (H. J. S. Smith). Show the following determinant expression for the continuant:

$$C(a_1,a_2,a_3,\dots,a_n) = \det \begin{pmatrix} a_1 & 1 & 0 & \cdots & 0 & 0 \\ -1 & a_2 & 1 & \cdots & 0 & 0 \\ 0 & -1 & a_3 & \cdots & 0 & 0 \\ \vdots & \vdots & \vdots & \ddots & \vdots & \vdots \\ 0 & 0 & \cdots & \cdots & a_{n-1} & 1 \\ 0 & 0 & \cdots & \cdots & -1 & a_n \end{pmatrix}.$$

4.2 Primes as sums of two squares

In this section we witness an application of continuants to the following classics.

Theorem 4.1 (Fermat, Gauss). *Any prime $p \equiv 1 \pmod 4$ can be represented in the form u^2+v^2, where $u,v \in \mathbb{Z}_{>0}$, and this representation is unique.*

Proof. Existence. Write our prime $p = 4r+1$ and, for each number $\mu \in \{2,3,\dots,2r\}$, run the Euclidean algorithm (4.1) for the fraction p/μ:

$$\frac{p}{\mu} = \frac{C(a_1,a_2,\dots,a_n)}{C(a_2,\dots,a_n)}.$$

Note that $2 < p/\mu < p/2$, so that $a_1 \ge 2$, and from $\gcd(p,\mu)=1$ we get

$$p = C(a_1,a_2,\dots,a_n) \quad \text{and} \quad \mu = C(a_2,\dots,a_n).$$

Observe that $p = C(a_n,a_{n-1},\dots,a_1)$ with $a_n \ge 2$ and take $\nu = C(a_{n-1},\dots,a_1)$. It follows then that

$$\frac{p}{\nu} = \frac{C(a_n,a_{n-1},\dots,a_1)}{C(a_{n-1},\dots,a_1)},$$

hence $\nu \in \{2, 3, \ldots, 2r\}$ as well. This defines an involution $\mu \leftrightarrow \nu$ on the set $A = \{2, 3, \ldots, 2r\}$, because applying the construction to ν we will get μ from it. The set A contains *odd* number of quantities, so that for at least one μ from the set we should have $\nu = \mu$. For such μ we thus obtain

$$\frac{C(a_1, a_2, \ldots, a_n)}{C(a_2, \ldots, a_n)} = \frac{p}{\mu} = \frac{C(a_n, a_{n-1}, \ldots, a_1)}{C(a_{n-1}, \ldots, a_1)}.$$

Because of the uniqueness of representation of p/μ by a continued fraction (in fact, the uniqueness of the Euclidean algorithm for the pair $p > \mu > 0$), the latter is only possible if $a_2 = a_{n-1}$, $a_3 = a_{n-2}$, $\ldots$, $a_n = a_1$; in other words, when $p = C(a_1, a_2, \ldots, a_2, a_1)$ has representation as a palindromic continuant. Consider now two possibilities: n is odd, and n is even.

If $n = 2s - 1$ for some $s \geq 2$, then $p = C(a_1, \ldots, a_{s-1}, a_s, a_{s-1}, \ldots, a_1)$, and Lemma 4.4 implies

$$\begin{aligned} p &= C(a_1, \ldots, a_{s-1}, a_s)\, C(a_{s-1}, \ldots, a_1) + C(a_1, \ldots, a_{s-1})\, C(a_{s-2}, \ldots, a_1) \\ &= \big(C(a_1, \ldots, a_{s-1}, a_s) + C(a_1, \ldots, a_{s-2})\big)\, C(a_1, \ldots, a_{s-1}) \end{aligned}$$

meaning that the prime p is divisible by the integer $C(a_1, \ldots, a_{s-1}) \geq a_1 \geq 2$, a contradiction.

If $n = 2s$ for some $s \geq 1$, then $p = C(a_1, \ldots, a_s, a_s, \ldots, a_1)$, and Lemma 4.4 implies

$$\begin{aligned} p &= C(a_1, \ldots, a_s)\, C(a_s, \ldots, a_1) + C(a_1, \ldots, a_{s-1})\, C(a_{s-1}, \ldots, a_1) \\ &= C(a_1, \ldots, a_s)^2 + C(a_1, \ldots, a_{s-1})^2, \end{aligned}$$

the required representation $p = u^2 + v^2$. □

Exercise 4.2. Let $p = 4r + 1$ be a prime. Then there are exactly $2r$ distinct representation of p as continuants $C(a_1, \ldots, a_n)$ with the first and last entries $a_1, a_n \geq 2$.

Hint. Use *distinct* representations coming from

$$\frac{p}{\mu} = \frac{C(a_1, a_2, \ldots, a_n)}{C(a_2, \ldots, a_n)}$$

when $\mu \in \{1, 2, \ldots, 2r\}$. □

Lemma 4.5. *For* $n = 2, 3, \ldots,$

$$\begin{aligned} &C(a_1, a_2, \ldots, a_{n-1}, a_n)\, C(a_2, \ldots, a_{n-1}) \\ &\qquad - C(a_1, a_2, \ldots, a_{n-1})\, C(a_2, \ldots, a_{n-1}, a_n) = (-1)^n. \end{aligned}$$

Proof. Simply compute the determinants on the both sides of (4.4). □

Theorem 4.2 (Euler). *Given a prime $p \equiv 1 \pmod 4$, there is a solution x_0 to the equation $x^2 \equiv -1 \pmod p$ with $1 \le x_0 < p/2$. In addition, the two different residues $\pm x_0 \pmod p$ exhaust all solutions to the equation.*

Proof. We use the palindromic representation $p = C(a_1, \dots, a_s, a_s, \dots, a_1)$ with $a_1 \ge 2$ found in the proof of Theorem 4.1. Define $x_0 = C(a_2, \dots, a_s, a_s, \dots, a_2, a_1)$, so that

$$\frac{p}{x_0} = \frac{C(a_1, \dots, a_s, a_s, \dots, a_1)}{C(a_2, \dots, a_s, a_s, \dots, a_2, a_1)} > a_1 \ge 2$$

and $1 \le x_0 < p/2$. It follows then from Lemma 4.5 applied with $n = 2s$ and Lemma 4.4 (a) that

$$\begin{aligned} 1 = (-1)^{2s} &= C(a_1, \dots, a_s, a_s, \dots, a_1)\, C(a_2, \dots, a_s, a_s, \dots, a_2) \\ &\quad - C(a_1, \dots, a_s, a_s, \dots, a_2)\, C(a_2, \dots, a_s, a_s, \dots, a_1) \\ &= pC(a_2, \dots, a_s, a_s, \dots, a_2) - x_0^2. \end{aligned}$$

Restricting this equality modulo p leads to $x_0^2 \equiv -1 \pmod p$. Clearly, both $x_0 \pmod p$ and $-x_0 \pmod p$ are solutions to $x^2 \equiv -1 \pmod p$, and the quadratic equation does not possess more than two solutions in the field $\mathbb{F}_p = \mathbb{Z}/p\mathbb{Z}$. □

Proof of Theorem 4.1. *Uniqueness.* Assume that there are two representations $p = u^2 + v^2 = u'^2 + v'^2$ of the given prime $p \equiv 1 \pmod 4$, with $u < v$ and $u' < v'$. Run the Euclidean algorithm on the rational numbers $v/u > 1$ and $v'/u' > 1$ to get the corresponding representations

$$\frac{v}{u} = \frac{C(a_1, a_2, \dots, a_s)}{C(a_2, \dots, a_s)} \quad \text{and} \quad \frac{v'}{u'} = \frac{C(a'_1, a'_2, \dots, a'_t)}{C(a'_2, \dots, a'_t)},$$

with $a_s \ge 2$ and $a'_t \ge 2$. From Lemma 4.4 we deduce that

$$\begin{aligned} p &= v^2 + u^2 \\ &= C(a_s, \dots, a_2, a_1)\, C(a_1, a_2, \dots, a_s) + C(a_s, \dots, a_2)\, C(a_2, \dots, a_s) \\ &= C(a_s, \dots, a_2, a_1, a_1, a_2, \dots, a_s) \end{aligned}$$

and, similarly,

$$p = v'^2 + u'^2 = C(a'_t, \dots, a'_2, a'_1, a'_1, a'_2, \dots, a'_t).$$

By the proof of Theorem 4.2 we know that both $C(a_{s-1}, \dots, a_1, a_1, \dots, a_{s-1}, a_s)$ and $C(a'_{t-1}, \dots, a'_1, a'_1, \dots, a'_{t-1}, a'_t)$ are solutions to $x^2 \equiv -1 \pmod p$ in the range $1 \le x < p/2$, so that they must coinside:

$$\mu = C(a_{s-1}, \dots, a_1, a_1, \dots, a_{s-1}, a_s) = C(a'_{t-1}, \dots, a'_1, a'_1, \dots, a'_{t-1}, a'_t).$$

It follows then from the uniqueness of the Euclidean algorithm for the pair $p > \mu > 0$ (equivalently, of the continued fraction for p/μ) that in the equality

$$\frac{p}{\mu} = \frac{C(a_s, a_{s-1}, \dots, a_1, a_1, \dots, a_{s-1}, a_s)}{C(a_{s-1}, \dots, a_1, a_1, \dots, a_{s-1}, a_s)} = \frac{C(a'_t, a'_{t-1}, \dots, a'_1, a'_1, \dots, a'_{t-1}, a'_t)}{C(a'_{t-1}, \dots, a'_1, a'_1, \dots, a'_{t-1}, a'_t)}$$

we have $t = s$ and $a'_i = a_i$ for all i. □

4.3 Continued fraction of a real number

It is now a good moment to introduce a compact notation for the finite continued fraction in (4.2):

$$[a_1, a_2, \dots, a_n] = a_1 + \cfrac{1}{a_2 + \cfrac{1}{a_3 + \ddots + \cfrac{1}{a_n}}} = \frac{C(a_1, a_2, \dots, a_n)}{C(a_2, \dots, a_n)}.$$

This is clearly a rational function of variables $a_1, a_2, \dots, a_n$ involved. At the same time, the expression originates from applying the Euclidean algorithm to a rational number $\alpha = a/b$ as in Section 4.1. The procedure can be alternatively interpreted as follows: take $a_1 = \lfloor \alpha \rfloor$ and, if α is not an integer, then write it in the form $\alpha = a_1 + 1/\alpha_2$, where $\alpha_2 > 1$ is again a rational number. Inductively, we choose $a_n = \lfloor \alpha_n \rfloor$ and $\alpha_n = a_n + 1/\alpha_{n+1}$ with $\alpha_{n+1} > 1$ if α_n is not an integer. The procedure terminates at some step (that is, eventually we get an integer $\alpha_n = a_n > 1$), so that $\alpha = [a_1, \dots, a_n]$. If we discard the condition $a_n > 1$ for $n \ge 1$ then the number α can be also represented as $\alpha = [a_1, \dots, a_n - 1, 1]$. This fact is sometimes useful for manipulating the parity of a particular length of a finite continued fraction.

The recursive algorithm above extends to the case of an *irrational* number α with no trouble; however, at each step we obtain irrational $\alpha_n > 1$, so that the continued fraction cannot be finite. We will use the notation

$$\alpha = [a_1, a_2, \dots, a_n, \dots]$$

for this infinite case. Observe that the procedure implies that $\alpha = \alpha_1$ and

$$\alpha = [a_1, a_2, \dots, a_n, \alpha_{n+1}]$$

for each $n \ge 0$, as well as $a_n < \alpha_n < a_n + 1$ for $n \ge 1$ and $a_n \ge 1$ for $n \ge 2$. Furthermore, by truncating the infinite continued fraction at the

nth step, we get the pair of relatively primes integers $p_n = C(a_1, a_2, \dots, a_n)$, $q_n = C(a_2, \dots, a_n)$ such that $[a_1, a_2, \dots, a_n] = p_n/q_n$. Here $1 = q_1 < q_2 < \cdots < q_n < \cdots$ because of the recursive formulae for q_n. The fraction $p_n/q_n = [a_1, a_2, \dots, a_n]$ is called the *n th (principal) convergent* of α, while a_n is the *n th partial quotient.*

Lemma 4.6. *For any $n \geq 2$,*

$$\frac{p_n}{q_n} - \frac{p_{n-1}}{q_{n-1}} = \frac{(-1)^n}{q_n q_{n-1}}.$$

Furthermore, for $n \geq 3$,

$$\frac{p_n}{q_n} - \frac{p_{n-2}}{q_{n-2}} = \frac{(-1)^{n-1} a_n}{q_n q_{n-2}}.$$

Proof. (a) Write the equality of Lemma 4.5 as $p_n q_{n-1} - p_{n-1} q_n = (-1)^n$ and divide both sides by $q_n q_{n-1}$.

(b) Similar to the proof of Lemma 4.3 we obtain

$$\begin{pmatrix} p_n & p_{n-2} \\ q_n & q_{n-2} \end{pmatrix} = \begin{pmatrix} a_1 & 1 \\ 1 & 0 \end{pmatrix} \begin{pmatrix} a_2 & 1 \\ 1 & 0 \end{pmatrix} \cdots \begin{pmatrix} a_{n-2} & 1 \\ 1 & 0 \end{pmatrix} \begin{pmatrix} a_{n-1} a_n + 1 & a_n \\ 1 & 0 \end{pmatrix}.$$

Passing to the determinant we arrive at $p_n q_{n-2} - p_{n-2} q_n = (-1)^{n-1} a_n$. Finally, divide both sides by $q_n q_{n-2}$. □

The second equality of Lemma 4.6 implies the following.

Lemma 4.7. *If $a_2, a_3, \dots$ are positive (not necessarily integer!) numbers then the sequence p_n/q_n restricted to odd n is strictly increasing, while restricted to even n it is strictly decreasing.*

Lemma 4.8. *For $n \geq 0$, we have the equalities*

$$q_{n+1}\alpha - p_{n+1} = \frac{(-1)^n}{\alpha_{n+2} q_{n+1} + q_n} \quad \textit{and} \quad q_n \alpha - p_n = \frac{(-1)^{n+1} \alpha_{n+2}}{\alpha_{n+2} q_{n+1} + q_n}.$$

Proof. Write the equalities in Lemma 4.6, after a shift, in the form

$$\frac{p_{n+2}}{q_{n+2}} - \frac{p_{n+1}}{q_{n+1}} = \frac{(-1)^n}{q_{n+2} q_{n+1}} = \frac{(-1)^n}{(a_{n+2} q_{n+1} + q_n) q_{n+1}}$$

for $n \geq 0$ and

$$\frac{p_{n+2}}{q_{n+2}} - \frac{p_n}{q_n} = \frac{(-1)^{n+1} a_{n+2}}{q_{n+2} q_n} = \frac{(-1)^{n+1} a_{n+2}}{(a_{n+2} q_{n+1} + q_n) q_n}$$

for $n \geq 1$. Now specify the variables $a_1, a_2, \dots, a_{n+1}, a_{n+2}$ involved to $a_1, a_2, \dots, a_{n+1}, \alpha_{n+2}$ corresponding to

$$\alpha = [a_1, a_2, \dots, a_n, \dots] = [a_1, a_2, \dots, a_n, a_{n+1}, \alpha_{n+2}]$$

and use the fact $p_{n+2}/q_{n+2} = \alpha$ after the specialisation. Finally, observe that the second formula is also true for $n = 0$ in view of $q_0 = 0$ and $p_0 = q_1 = 1$. □

Theorem 4.3 (monotonicity and estimation of convergents). *For odd n, the nth convergents of α form a strictly increasing sequence converging to α; for even n, the nth convergents of α form a strictly decreasing sequence converging to α. Furthermore, for $n \geq 1$,*

$$\frac{1}{2q_n q_{n+1}} < \frac{1}{q_n(q_n + q_{n+1})} < \left|\alpha - \frac{p_n}{q_n}\right| < \frac{1}{q_n q_{n+1}} \leq \frac{1}{a_{n+1}q_n^2}.$$

Proof. By making a reference to Lemma 4.7, we only need to explain the estimates. Because α is always located between two consecutive convergents p_n/q_n and p_{n+1}/q_{n+1}, we deduce that

$$\left|\alpha - \frac{p_n}{q_n}\right| < \left|\frac{p_{n+1}}{q_{n+1}} - \frac{p_n}{q_n}\right| = \frac{1}{q_{n+1}q_n};$$

in addition, $q_{n+1} = a_{n+1}q_n + q_{n-1} \geq a_{n+1}q_n$ for $n \geq 1$. Similarly, by locating α with respect to p_n/q_n and p_{n+2}/q_{n+2} we obtain

$$\left|\alpha - \frac{p_n}{q_n}\right| > \left|\frac{p_{n+2}}{q_{n+2}} - \frac{p_n}{q_n}\right| = \frac{a_{n+2}}{q_{n+2}q_n} = \frac{a_{n+2}}{(a_{n+2}q_{n+1} + q_n)q_n} \geq \frac{1}{(q_{n+1} + q_n)q_n},$$

since $a_{n+2} \geq 1$. □

Since $q_{n+1} > q_n$, we conclude that the principal convergents p_n/q_n satisfy the inequality

$$\left|\alpha - \frac{p}{q}\right| < \frac{1}{q^2}.$$

Exercise 4.3. Prove that

$$\alpha = a_1 + \sum_{n=2}^{\infty} \frac{(-1)^n}{q_n q_{n-1}}.$$

Hint. Use Lemma 4.6 to show that

$$\frac{p_n}{q_n} = a_1 + \sum_{j=2}^{n} \frac{(-1)^j}{q_j q_{j-1}}$$

and then apply the convergence of p_n/q_n to α as $n \to \infty$. □

Exercise 4.4. Define $S_0 = 2$ and $S_{n+1} = S_n^2 - S_n + 1$ for $n \geq 0$; this is Sylvester's sequence.

(a) Using Exercise 4.3 show that the partial quotients in the continued fraction expansion of

$$C = \sum_{n=0}^{\infty} \frac{(-1)^n}{S_n - 1}$$

are all squares.

(b) Prove that

$$C' = \sum_{n=0}^{\infty} \frac{(-1)^n}{S_n}$$

has a continued fraction $[a_1, a_2, \ldots, a_n, \ldots]$ where for $n \geq 3$ each $a_n/2$ is a square.

(c) Finally, show that $2C = C' + 1$.

(The number $C = 0.64341054628\ldots$ is sometimes called *Cahen's constant.*)

4.4 A taste of diophantine approximation

Lemma 4.9. *Let p_{n-1}/q_{n-1} and p_n/q_n be two successive convergents of an irrational number $\alpha = [a_1, a_2, \ldots]$. Then at least one of these fractions satisfies the inequality*

$$\left|\alpha - \frac{p}{q}\right| < \frac{1}{2q^2}.$$

Proof. Assume, to get a contradiction, that

$$\left|\alpha - \frac{p_{n-1}}{q_{n-1}}\right| \geq \frac{1}{2q_{n-1}^2} \quad \text{and} \quad \left|\alpha - \frac{p_n}{q_n}\right| \geq \frac{1}{2q_n^2}.$$

Using the fact that α lies between p_{n-1}/q_{n-1} and p_n/q_n (Theorem 4.3) we obtain

$$\frac{1}{q_{n-1}q_n} = \left|\frac{p_{n-1}}{q_{n-1}} - \frac{p_n}{q_n}\right| = \left|\alpha - \frac{p_{n-1}}{q_{n-1}}\right| + \left|\alpha - \frac{p_n}{q_n}\right| \geq \frac{1}{2q_{n-1}^2} + \frac{1}{2q_n^2}.$$

This contradicts the inequality $xy < (x^2 + y^2)/2$ for $x > y > 0$, applied with $x = 1/q_{n-1}$ and $y = 1/q_n$. □

Our next statement shows that, in a certain sense, the converse of Lemma 4.9 holds as well.

Theorem 4.4 (Legendre). *Let p and q be coprime integers, $q > 0$, and let*

$$\left|\alpha - \frac{p}{q}\right| < \frac{1}{2q^2}.$$

Then p/q is a convergent of α.

Proof. Write the continued fraction expansion of the rational number p/q: $p/q = [a_1, a_2, \ldots, a_n]$. Let p_{n-1}/q_{n-1} and $p/q = p_n/q_n$ be the last two convergents of this expansion, where we assume that both α and p_{n-1}/q_{n-1} are simultaneously greater or smaller than the number p/q (if this does not happen then we replace the continued fraction with $p/q = [a_1, \ldots, a_{n-1}, a_n - 1, 1]$). Consider the number

$$\beta = -\frac{p_{n-1} - \alpha q_{n-1}}{p_n - \alpha q_n}, \tag{4.5}$$

for which we have

$$\begin{aligned} \left|\beta + \frac{q_{n-1}}{q_n}\right| &= \left|\frac{p_{n-1} - \alpha q_{n-1}}{\alpha q_n - p_n} + \frac{q_{n-1}}{q_n}\right| \\ &= \frac{1}{q_n^2|\alpha - p_n/q_n|} = \frac{1}{q^2|\alpha - p/q|} > 2, \end{aligned}$$

implying

$$|\beta| \geq \left|\beta + \frac{q_{n-1}}{q_n}\right| - \frac{q_{n-1}}{q_n} > 2 - \frac{q_{n-1}}{q_n} \geq 1. \tag{4.6}$$

Comparing the latter inequality with (4.5) we deduce that

$$|p_{n-1} - \alpha q_{n-1}| > |p_n - \alpha q_n|;$$

hence

$$\left|\alpha - \frac{p_{n-1}}{q_{n-1}}\right| > \left|\alpha - \frac{p_n}{q_n}\right|.$$

However, the numbers α and p_{n-1}/q_{n-1} are both either greater or smaller than p/q, that is, α lies between p_{n-1}/q_{n-1} and $p/q = p_n/q_n$. But then $\beta > 0$ in accordance with (4.5), and so $\beta > 1$ by (4.6).

Let $[a_{n+1}, a_{n+2}, \ldots]$ be the continued fraction of β; we have $a_{n+1} \geq 1$ in view of $\beta > 1$. Then

$$[a_1, \ldots, a_n, a_{n+1}, \ldots] = [a_1, \ldots, a_n, \beta] = \alpha;$$

in other words, $p/q = p_n/q_n$ is indeed a convergent of α. □

4.5 Equivalent numbers

The set of matrices

$$\gamma = \begin{pmatrix} a & b \\ c & d \end{pmatrix}$$

with integer entries a, b, c, d and determinant ± 1 (that is, $ad-bc = 1$ or -1) is a multiplicative group with identity (*neutral*) element

$$E = \begin{pmatrix} 1 & 0 \\ 0 & 1 \end{pmatrix}.$$

Indeed, the product of any two such matrices and the inverse of such a matrix again has integer entries and determinant equal to ± 1. This group is known as the *general linear group* (over the ring $\mathbb{Z}$) and is denoted by $GL_2(\mathbb{Z})$; in what follows we reserve the notation Γ for this group.

For an irrational number α, the action of an element $\gamma \in \Gamma$ is defined by the rule

$$\gamma\alpha = \frac{a\alpha + b}{c\alpha + d}.$$

Exercise 4.5. Show that the action is well defined, namely, that $E\alpha = \alpha$ and $\gamma(\delta\alpha) = (\gamma\delta)\alpha$ for all $\gamma, \delta \in \Gamma$.

We say that two irrational numbers α and β are *equivalent* if $\gamma\alpha = \beta$ for some $\gamma \in \Gamma$.

Exercise 4.6. Show that this relation is indeed an equivalence.

For an irrational number α we have the representation

$$\alpha = [a_1, \dots, a_n, \alpha_{n+1}] = \frac{p_n\alpha_{n+1} + p_{n-1}}{q_n\alpha_{n+1} + q_{n-1}},$$

in accordance with Lemma 4.3. Define the *nth continued transformation* of the number α by the equality

$$\gamma_n = \begin{pmatrix} p_n & p_{n-1} \\ q_n & q_{n-1} \end{pmatrix} = \prod_{j=1}^{n} \begin{pmatrix} a_j & 1 \\ 1 & 0 \end{pmatrix};$$

note that $\gamma_n \in \Gamma$ from computing the determinants in the latter product. Then $\alpha = \gamma_n\alpha_{n+1}$, and hence α is equivalent to α_n for any $n \geq 1$. In other words, all complete quotients α_n, $n = 1, 2, \dots$, are equivalent to each other.

The following theorem characterises the situation considered in this example.

Theorem 4.5. *Let* $\alpha, \beta \in \mathbb{R} \setminus \mathbb{Q}$ *and*

$$\alpha = \gamma\beta = \frac{a\beta + b}{c\beta + d} \quad \textit{for some } \gamma = \begin{pmatrix} a & b \\ c & d \end{pmatrix} \in \Gamma.$$

Assume that $\beta > 1$ *and* $c > d > 0$. *Then* b/d *and* a/c *are two consecutive convergents of* α, *say,* p_{n-1}/q_{n-1} *and* p_n/q_n; *furthermore,* $\beta = \alpha_{n+1}$.

Proof. Note that a and c are relatively prime since $ad - bc = \pm 1$. Write a/c as the finite continued fraction

$$\frac{a}{c} = [a_1, \dots, a_{n-1}, a_n] = \frac{p_n}{q_n}, \quad a_n > 1,$$

where $a = p_n$ and $c = q_n$. Increasing by 1, if required, the length of the continued fraction for a/c (namely, replacing a_n with $a_n - 1, 1$ in its record), we obtain the equality

$$p_n q_{n-1} - q_n p_{n-1} = \varepsilon,$$

where $\varepsilon = ad - bc$. Then

$$ad - bc = p_n d - q_n b = \varepsilon$$

and comparing the two equations we deduce that

$$p_n(d - q_{n-1}) = q_n(b - p_{n-1}). \tag{4.7}$$

Since p_n and q_n are coprime, we conclude from (4.7) that q_n divides $d - q_{n-1}$; but $q_{n-1} \leq q_n$ and $0 < d < c = q_n$, that is, $|d - q_{n-1}| < q_n$, so that $d - q_{n-1} = 0$. Then (4.7) implies that $b - p_{n-1} = 0$. Therefore,

$$\alpha = \frac{a\beta + b}{c\beta + d} = \frac{p_n\beta + p_{n-1}}{q_n\beta + q_{n-1}} = [a_1, \dots, a_n, \beta].$$

By the hypothesis $\beta > 1$, so the resulting expression is the continued fraction representing the number α and we have $\beta = \alpha_{n+1}$. This means that b/d and a/c are consecutive convergents of α. □

Theorem 4.6 (Serret). *Two numbers $\alpha, \beta \in \mathbb{R} \setminus \mathbb{Q}$ are equivalent if and only if there exist integers $n, m \geq 1$ such that $\alpha_n = \beta_m$. In other words, α and β are equivalent if and only if their continued fractions are*

$$\alpha = [a_1, a_2, \dots] \quad \textit{and} \quad \beta = [b_1, b_2, \dots]$$

and $a_n = b_{n+l}$ for some $l \in \mathbb{Z}$ and all $n \geq N$.

Proof. First assume that for some $n, m \geq 1$ we have $\alpha_n = \beta_m$, that is,

$$\alpha = [a_1, \dots, a_{n-1}, \alpha_n], \quad \beta = [b_1, \dots, b_{m-1}, \beta_m],$$

and that $\alpha_n = \beta_m$. Since α is equivalent to α_n and β is equivalent to β_m (as we have already seen), we conclude that α and β are equivalent.

Conversely, suppose that α and β are equivalent, that is,

$$\beta = \frac{a\alpha + b}{c\alpha + d} = \gamma\alpha, \quad ad - bc = \pm 1.$$

Changing, if necessary, the signs of all entries of γ to their opposites, we may assume that $c\alpha+d>0$. Let γ_{n-1} be the $(n-1)$th continued transformation of α; thus $\alpha=\gamma_{n-1}\alpha_n$. Then $\beta=\gamma\gamma_{n-1}\alpha_n$, and

$$\gamma\gamma_{n-1}=\begin{pmatrix} ap_{n-1}+bq_{n-1} & ap_{n-2}+bq_{n-2} \\ cp_{n-1}+dq_{n-1} & cp_{n-2}+dq_{n-2} \end{pmatrix}=\begin{pmatrix} a' & b' \\ c' & d' \end{pmatrix}.$$

We have

$$\begin{aligned} cp_{n-1}+dq_{n-1}&=q_{n-1}\left(c\frac{p_{n-1}}{q_{n-1}}+d\right)=c', \\ cp_{n-2}+dq_{n-2}&=q_{n-2}\left(c\frac{p_{n-2}}{q_{n-2}}+d\right)=d'. \end{aligned} \tag{4.8}$$

Take n large enough that both p_{n-2}/q_{n-2} and p_{n-1}/q_{n-1} are close to α; the parity of n is chosen depending on the sign of c to have

$$c\frac{p_{n-2}}{q_{n-2}}<c\alpha<c\frac{p_{n-1}}{q_{n-1}}.$$

Then $c'>0$, $d'>0$ and, in addition, $\alpha_n>1$; from (4.8) we have $c'>d'$ as $q_{n-2}<q_{n-1}$. Thus, all the conditions of Theorem 4.5 are fulfilled, and we conclude that $\alpha_n=\beta_m$ for some m. This completes our proof of the theorem. □

4.6 Continued fraction of a quadratic irrational

Let d be a positive integer. It can be seen that the set $\{x+y\sqrt{d} : x,y\in\mathbb{Q}\}$ forms a field. In what follows, we assume that this field does not coincide with $\mathbb{Q}$, in other words, that d is not a perfect square. Moreover, without loss of generality we may assume that the number d is square-free (that is, d is not divisible by a square >1).

Note that 1 and $\sqrt{d}$ are linearly independent over $\mathbb{Q}$ (otherwise $\sqrt{d}$ would be rational). This implies that each element of the field possesses a *unique* representation in the form $x+y\sqrt{d}$ with $x,y\in\mathbb{Q}$. Let this field be denoted by $\mathbb{Q}(\sqrt{d})$ and define the *conjugate* of a number $\alpha=x+y\sqrt{d}$ to be $\overline{\alpha}=x-y\sqrt{d}$.

Exercise 4.7. Verify that

$$\overline{\alpha+\beta}=\overline{\alpha}+\overline{\beta} \quad \text{and} \quad \overline{\alpha\beta}=\overline{\alpha}\overline{\beta}.$$

Now define the *trace* and the *norm* of a number $\alpha\in\mathbb{Q}(\sqrt{d})$ by

$$\mathrm{Tr}(\alpha)=\alpha+\overline{\alpha}=2x\in\mathbb{Q}, \quad \mathrm{N}(\alpha)=\alpha\overline{\alpha}=x^2-dy^2\in\mathbb{Q}.$$

Then α and its conjugate $\overline{\alpha}$ are the roots of the quadratic polynomial

$$(x-\alpha)(x-\overline{\alpha}) = x^2 - \mathrm{Tr}(\alpha)\,x + \mathrm{N}(\alpha)$$

with rational coefficients; this characterises α as a *quadratic irrational.* Thus, a defining equation for the quadratic irrational α can be written in the form

$$\alpha^2 - 2x\alpha + (x^2 - dy^2) = 0;$$

taking $x^2 - dy^2 = c/a$ and $-2x = b/a$, where a, b, c are coprime integers and $a > 0$, we can represent the quadratic equation as

$$a\alpha^2 + b\alpha + c = 0$$

with coprime integers a, b, c, $a > 0$. Such a, b, c are determined by α uniquely. Finally, define the *discriminant* of a quadratic irrational α by the formula

$$D(\alpha) = b^2 - 4ac = 4a^2y^2d.$$

Since α is a real irrational number, we have $D(\alpha) > 0$.

We shall call α a *reduced* quadratic irrational if $\alpha > 1$ and $-1 < \overline{\alpha} < 0$ (equivalently, $-1/\overline{\alpha} > 1$).

Exercise 4.8. If α is a reduced quadratic irrational, show that $-1/\overline{\alpha}$ is reduced as well.

Theorem 4.7. *For a given positive integer D, there exist at most finitely many reduced elements of the field $\mathbb{Q}(\sqrt{d})$ whose discriminant is equal to D.*

Proof. Let α be a reduced number having discriminant $D(\alpha) = D$. Then

$$\alpha = \frac{-b + \varepsilon\sqrt{D}}{2a} > 1 \quad \text{and} \quad -1 < \frac{-b - \varepsilon\sqrt{D}}{2a} < 0, \tag{4.9}$$

where $\varepsilon = 1$ or -1. If $\varepsilon = -1$ then we obtain $\alpha < 0$, which is impossible. Therefore $\varepsilon = 1$ and, in accordance with $a > 0$ and (4.9),

$$b + \sqrt{D} < 2a < -b + \sqrt{D}. \tag{4.10}$$

This means that $b < 0$; furthermore, the second inequality in (4.9) implies $-b < \sqrt{D}$. From these bounds on $|b|$ we conclude that there are finitely many possibilities for the quantity b to satisfy the inequalities (4.9). In turn, the inequality (4.10) retains only finitely many possibilities for the quantity $a > 0$ as well. Finally, the quantity $c \in \mathbb{Z}$ (if it exists) is subject to the relation $b^2 - 4ac = D$, and hence it is determined uniquely by the three quantities D, a and b. □

Lemma 4.10. *If α has discriminant $D > 0$ and β is equivalent to α then β has the same discriminant D.*

Proof. For $\alpha = x + y\sqrt{d}$ write

$$\alpha = \frac{A\beta + B}{E\beta + F}, \quad AF - BE = \pm 1.$$

Then

$$a\left(\frac{A\beta + B}{E\beta + F}\right)^2 + b\frac{A\beta + B}{E\beta + F} + c = 0$$

is equivalent to the quadratic equation

$$\begin{aligned} &a(A\beta + B)^2 + b(A\beta + B)(E\beta + F) + c(E\beta + F)^2 \\ &\quad = (aA^2 + bAE + cE^2)\beta^2 + (2aAB + bAF + bBE + 2cEF)\beta \\ &\quad\quad + (aB^2 + bBF + cF^2) = 0 \end{aligned}$$

whose discriminant is equal to

$$\begin{aligned} &(2aAB + bAF + bBE + 2cEF)^2 \\ &\quad\quad - 4(aA^2 + bAE + cE^2)(aB^2 + bBF + cF^2) \\ &\quad = b^2 - 4ac = D(\alpha), \end{aligned}$$

and whose coefficients are coprime. (If there is a common multiple of the coefficients then the inverse transformation

$$\beta = \frac{F\alpha - B}{-E\tau + A}$$

leads to the original quadratic equation for α, with coefficients a, b, c having the same common multiple.) □

Theorem 4.8. *Let α be a real quadratic irrational number. Then*

(i) *the number α_n, $n \geq 1$, in the continued fraction*

$$\alpha = [a_1, \ldots, a_{n-1}, \alpha_n]$$

has the same discriminant as α;

(ii) *if α is a reduced number then α_n is reduced for any $n \geq 1$ as well; and*

(iii) *if α is not necessarily reduced then α_n is reduced for all n sufficiently large.*

Proof. Claim (i) follows from Lemma 4.10. Moreover, the defining procedure of the continued fraction for α implies $\alpha_n > 1$ for all $n \geq 1$.

(ii) If α is reduced then $\alpha = a + 1/\beta$ for an integer $a \geq 1$ and a real $\beta > 1$; this implies $-1/\overline{\beta} = a - \overline{\alpha} > 1$, since $a \geq 1$ and $\overline{\alpha} < 0$. Therefore β is a reduced number as well.

(iii) From

$$\alpha = \gamma_n \alpha_{n+1} = \frac{p_n \alpha_{n+1} + p_{n-1}}{q_n \alpha_{n+1} + q_{n-1}}$$

we have

$$\alpha_{n+1} = -\frac{q_{n-1}\alpha - p_{n-1}}{q_n \alpha - p_n} \tag{4.11}$$

implying

$$\overline{\alpha}_{n+1} = -\frac{q_{n-1}\overline{\alpha} - p_{n-1}}{q_n \overline{\alpha} - p_n} = -\frac{q_{n-1}}{q_n} \frac{\overline{\alpha} - p_{n-1}/q_{n-1}}{\overline{\alpha} - p_n/q_n}. \tag{4.12}$$

Eventually the fractions p_{n-2}/q_{n-2} and p_{n-1}/q_{n-1} become close to α, so that both the numerator and denominator of the last fraction are close to $\overline{\alpha} - \alpha$; in particular, they have the same sign. Consequently, $\overline{\alpha}_n < 0$. Furthermore,

$$\begin{aligned} \frac{\overline{\alpha} - p_{n-1}/q_{n-1}}{\overline{\alpha} - p_n/q_n} &= 1 + \frac{p_n/q_n - p_{n-1}/q_{n-1}}{\overline{\alpha} - p_n/q_n} \\ &= 1 + \frac{(-1)^n}{q_n q_{n-1}(\overline{\alpha} - p_n/q_n)}. \end{aligned}$$

Continuing (4.12) we find that

$$\overline{\alpha}_{n+1} + 1 = \frac{1}{q_n}\left(q_n - q_{n-1} - \frac{(-1)^n}{q_n(\overline{\alpha} - p_n/q_n)}\right).$$

The expression

$$\frac{1}{q_n(\overline{\alpha} - p_n/q_n)}$$

tends to 0 as $n \to \infty$, hence its absolute value is less than 1 for all n sufficiently large. This implies $\overline{\alpha}_{n+1} + 1 > 0$ and finishes our proof of claim (iii). □

As a side application, we may iterate (4.11) to derive

Exercise 4.9 (distance formula). Show that for $n \geq 1$

$$\alpha_2 \cdots \alpha_n \alpha_{n+1} = \frac{(-1)^{n+1}}{p_n - q_n \alpha},$$

with our previous convention $p_0 = 1$, $q_0 = 0$.

Hint. Iterate equality (4.11). □

It turns out that one may usefully think of $\left|\log|p_n - q_n\alpha|\right|$ as measuring a weighted distance that the continued fraction has traversed in moving from α to α_{n+1}.

4.7 Euler–Lagrange theorem

Let α be a real irrational number. We say that its continued fraction

$$[a_1, a_2, a_3, \ldots]$$

is *periodic* if there exists an integer k such that $a_{n+k} = a_n$ for all n sufficiently large and *purely periodic* if $a_{n+k} = a_n$ for all $n \geq 1$; we call k the *primitive period* if it is the smallest positive integer with the above property.

The following standard notation is used for periodic continued fractions:

$$[a_1, \ldots, a_r, \overline{a_{r+1}, \ldots, a_{r+k}}\,],$$

where the vinculum (overbar) denotes the periodic repetition of the corresponding part. A continued fraction is purely periodic iff it can be written in the form $[\,\overline{a_1, \ldots, a_k}\,]$.

Lemma 4.11. *Let α be a reduced quadratic irrational and a an integer. Write $\alpha = a + 1/\beta$. Then β is reduced iff $a < \alpha < a + 1$, that is, iff $a = \lfloor \alpha \rfloor$.*

Proof. If $a = \lfloor \alpha \rfloor$ then $\beta > 1$ and $-1/\overline{\beta} = a - \overline{\alpha} > a = \lfloor \alpha \rfloor \geq 1$, hence β is reduced.

Conversely, if $\alpha < a$ then $\beta < 0$, and if $a + 1 < \alpha$ then $\beta < 1$; thus β cannot be reduced if $a \neq \lfloor \alpha \rfloor$. □

We point out that the relation between α and β in Lemma 4.11 determines one of these numbers in terms of the other. Indeed,

$$-1/\overline{\beta} = a + \frac{1}{-1/\overline{\alpha}},$$

which implies that $a = \lfloor -1/\overline{\beta} \rfloor$. Moreover, $\overline{\alpha}$ (and hence α itself) is uniquely determined by $\overline{\beta}$ or, hence, by β.

We now come to a central result characterising quadratic irrationals in terms of the periodicity of their continued fractions. Recall, in contrast, that the eventual periodicity of its base-b expansion characterises the rationality of the number. This points to the power of continued fraction representations over b-ary ones.

Theorem 4.9 (Euler–Lagrange theorem). *Let α be a real irrational number. The continued fraction for α is periodic iff α is a quadratic irrational. In the latter case, α is reduced iff its continued fraction is purely periodic.*

Proof. First assume that α is a quadratic irrational. By Theorem 4.8(iii) the corresponding tails α_n are reduced for all $n \geq n_0$, while Theorem 4.7, together with Lemma 4.10, implies the finiteness of the reduced numbers that are equivalent to α. Therefore, for some $n \geq n_0$ and $k > 1$ we have $\alpha_n = \alpha_{n+k}$. This immediately implies the periodicity of the continued fraction. Furthermore, assume that α itself is reduced; by part (ii) of Theorem 4.8 all the α_n are reduced as well. As we already know, $\alpha_n = \alpha_{n+k}$ for some n and $k \geq 1$. From Lemma 4.11 and the comment to it, we conclude that α_{n-1} is uniquely determined by α_n and hence that $\alpha_{n-1} = \alpha_{n+k-1}$. Applying this descent n times, we finally arrive at $\alpha = \alpha_1 = \alpha_{k+1}$; in other words, the continued fraction is purely periodic.

Conversely, if a continued fraction is purely periodic then it may be written as

$$\alpha = [\,\overline{a_1, \ldots, a_k}\,] = [a_1, \ldots, a_k, \alpha].$$

The relation $\alpha = \gamma_k \alpha$ implies that α is a root of a quadratic equation with integer coefficients, while by claim (iii) of Theorem 4.8 the number $\gamma_k^n \alpha = \alpha$ is reduced. In the case of a periodic continued fraction, we write

$$\alpha = [a_1, \ldots, a_r, \overline{a_{r+1}, \ldots, a_{r+k}}\,] = [a_1, \ldots, a_r, \alpha_{r+1}],$$

where the purely periodic continued fraction $\alpha_{r+1} = [\,\overline{a_{r+1}, \ldots, a_{r+k}}\,]$ is, by the above argument, a (reduced) quadratic irrational. Since α and α_{r+1} are equivalent, the number α is a quadratic irrational as well. □

Exercise 4.10. Show that if α is reduced and $\alpha = [\,\overline{a_1, \ldots, a_k}\,]$ then $-1/\overline{\alpha} = [\,\overline{a_k, a_{k-1}, \ldots, a_1}\,]$.

Exercise 4.11 (Perron's theorem). Let β be a real number. Show that β is the square root of a rational number > 1 iff there exist an integer $b_1 > 0$ and a finite (possibly empty) *palindromic* list of positive integers $b_2, \ldots, b_k$ such that $\beta = [b_1, \overline{b_2, \ldots, b_k, 2b_1}]$.

Sketch of solution. An equivalent way of saying that a list $b_2, b_3, \ldots, b_k$ is palindromic is that the matrix

$$\begin{pmatrix} a & b \\ c & d \end{pmatrix} = \begin{pmatrix} b_2 & 1 \\ 1 & 0 \end{pmatrix} \begin{pmatrix} b_3 & 1 \\ 1 & 0 \end{pmatrix} \cdots \begin{pmatrix} b_k & 1 \\ 1 & 0 \end{pmatrix}$$

is symmetric (that is, $b = c$). Writing

$$\begin{aligned}\beta &= [b_1, \overline{b_2, \dots, b_k, 2b_1}] = [b_1, b_2, \dots, b_k, b_1 + \beta] \\ &= b_1 + \frac{1}{[b_2, \dots, b_k, b_1 + \beta]} = b_1 + \frac{c(b_1 + \beta) + d}{a(b_1 + \beta) + b}\end{aligned}$$

we obtain a quadratic equation for β,

$$a\beta^2 + (b - c)\beta - b_1(ab_1 + b + c) = 0,$$

whose linear term vanishes iff $b = c$. □

Chapter notes

This chapter is reasonably elementary and also independent of the remaining contents; at the same time it provides us with a natural bridge between integer investigations (via an extension of the Euclidean algorithm) and diophantine questions (representation as sums of squares and rational approximations of real numbers). Continued fractions are a self-standing topic in number theory, with many books dedicated to them; one recommendation would be [14] which shares the style with this book.

The Fermat–Gauss theorem about representativeness of primes of the form $4k + 1$ as sums of two squares is a record keeper among the results with multiple different proofs available for; some of them can be found in [1, Chapter 4]. One more proof can be cheaply extracted from the equality of power series

$$\left(\sum_{m=-\infty}^{\infty} q^{m^2} \right)^2 = 1 + 4 \sum_{n=0}^{\infty} \frac{(-1)^n q^{2n+1}}{1 - q^{2n+1}}$$

highlighted at the end of Chapter 1; both sides represent a q-deformation of π.

Chapter 5

Dirichlet's theorem on primes in arithmetic progressions

5.1 Quadratic residues

For a positive integer m, we say that two integers a and b are congruent modulo m and write $a \equiv b \pmod{m}$ if their difference is divisible by m. All integers that are congruent to a particular number a modulo m form the residue class $a \pmod{m}$. Here are two basic properties of congruences you can think about.

Lemma 5.1. *Assume that $a \equiv b \pmod{m}$ and $c \equiv d \pmod{m}$. Then $a \pm c \equiv b \pm d \pmod{m}$ and $ac \equiv bd \pmod{m}$.*

Lemma 5.2. *Assume that $ac \equiv bc \pmod{m}$ and $(c, m) = 1$. Then $a \equiv b \pmod{m}$.*

Dirichlet's theorem whose proof we discuss in this chapter asserts that for a fixed pair l, m of two positive relatively prime integers, there are infinitely many primes $p \equiv l \pmod{m}$.

Lemma 5.3. *Let $m \geq 1$ and a, b be integers such that $(a, m) = 1$. Then all solutions x of the congruence equation $ax \equiv b \pmod{m}$ form a single residue class modulo m.*

Proof. If x_0 is a solution of the congruence $ax \equiv b \pmod{m}$ and $x_1 \equiv x_0 \pmod{m}$, then clearly $ax_1 \equiv b \pmod{m}$, so that x_1 is a solution as well. In the other direction, if x_0 and x_1 are two solutions of the congruence then, by subtracting, we get $a(x_1 - x_0) \equiv 0 \pmod{m}$, so that $x_1 \equiv x_0 \pmod{m}$. □

Euler's totient function $\varphi(m)$ assigns to each $m \geq 1$ the number of integers in the range $\{0, 1, \ldots, m-1\}$ (or their related residue classes modulo m), which are coprime with m. For example, $\varphi(1) = \varphi(2) = 1$ and $\varphi(p^k) = p^k - p^{k-1}$ for a prime p and $k \geq 1$.

Exercise 5.1. Show that $\varphi(m)$ is a multiplicative function and deduce the formula

$$\varphi(m) = m \prod_{p|m} \left(1 - \frac{1}{p}\right)$$

from this.

Theorem 5.1 (Euler's theorem). *Let m be a positive integer and a relatively prime to m. Then $a^{\varphi(m)} \equiv 1 \pmod{m}$.*

Proof. Consider the collection $\{r_1, \dots, r_n\} \subset \{0, 1, \dots, m-1\}$ of $n = \varphi(m)$ integers coprime to m. Then the collection $\{ar_1, \dots, ar_n\}$ represents *different* residue classes modulo m such that $(ar_j, m) = 1$ for all j. This means that $\{ar_1, \dots, ar_n\}$ is a permutation of $\{r_1, \dots, r_n\}$ modulo m; in particular, the products $\prod_{j=1}^n (ar_n)$ and $\prod_{j=1}^n r_j$ are congruent modulo m. By cancelling the latter product in $\prod_{j=1}^n (ar_n) \equiv \prod_{j=1}^n r_j \pmod{m}$ (with the help of Lemma 5.3) we arrive at the desired claim. □

Taking $m = p$ in the theorem we get Fermat's little theorem.

Exercise 5.2. Compute the product $\prod_{j=1}^n r_j \pmod{m}$ in the proof of Theorem 5.1.

Exercise 5.3. Prove Wilson's theorem: A number $m > 1$ is prime if and only if $(m-1)! \equiv -1 \pmod{m}$.

5.2 Infinitude of primes of the form $4n \pm 1$

The next two results are particular cases of Dirichlet's theorem whose proofs can be accomplished by elementary consideration.

Theorem 5.2. *There are infinitely many primes of the form $4n - 1$.*

Proof. Assume on the contrary that there are finitely many of them, $p_1, \dots, p_r$ say. Then at least one of the prime factors of $N = 4p_1 \cdots p_r - 1$ must be of this form (otherwise, the number N will be congruent to 1 modulo 4). On the other hand, that prime is on our finite list because $(N, p_j) = 1$ for all j; contradiction. □

Theorem 5.3. *There are infinitely many primes of the form $4n + 1$.*

Proof. First notice that the congruence $x^2 \equiv -1 \pmod{p}$ can only be solved in integers x for (odd!) primes p of the form $4n+1$ (and we construct those solutions x in Theorem 4.2). Indeed, if $p \equiv -1 \pmod 4$, so that $p = 4n-1$ for some n, then $x^{\varphi(p)} = x^{p-1} = (x^2)^{2n-1} \equiv (-1)^{2n-1} = -1 \pmod{p}$ in contradiction with Theorem 5.1.

The remaining part of the argument is similar to that in our proof of Theorem 5.2. Assume on the contrary that there are finitely many primes $p_1, \dots, p_r$ of the form $4n+1$. Then the least prime divisor of the odd number $N = (2p_1 \cdots p_r)^2 + 1$ has the same form and is not on the list; contradiction. $\square$

Exercise 5.4. Let p be prime, $p > 3$, and the congruence

$$x^2 + x + 1 \equiv 0 \pmod{p}$$

has a solution $x \in \mathbb{Z}$. Prove that p has the form $6n+1$. Deduce from this result that there are infinitely primes of this form.

Exercise 5.5. Let p be prime, $p > 5$, and the congruence

$$x^4 + x^3 + x^2 + x + 1 \equiv 0 \pmod{p}$$

has a solution $x \in \mathbb{Z}$. Prove that p has the form $10n+1$. Deduce from this result that there are infinitely primes of this form.

Exercise 5.6. Let p be prime, $p > 2$, and the congruence

$$x^4 + 1 \equiv 0 \pmod{p}$$

has a solution $x \in \mathbb{Z}$. Prove that p has the form $8n+1$. Deduce from this result that there are infinitely primes of this form.

Exercise 5.7. Let p be prime, $p > 2$, and the congruence

$$x^2 + 2 \equiv 0 \pmod{p}$$

has a solution $x \in \mathbb{Z}$. Prove that p is either of the form $8n+1$ or $8n+3$. Deduce from this result that there are infinitely primes of the form $8n+3$.

Hint. The results of this type are related to calculation of the Legendre symbol

$$\left(\frac{a}{p}\right) = \begin{cases} 0 & \text{if } a \equiv 0 \pmod{p}, \\ 1 & \text{if } x^2 \equiv a \pmod{p} \text{ is soluble in } x \in \mathbb{Z}, \\ -1 & \text{if } x^2 \equiv a \pmod{p} \text{ is not soluble in } x \in \mathbb{Z}, \end{cases}$$

where p is an odd prime and $a \in \mathbb{Z}$ is arbitrary. The statement in Exercise 5.7 is equivalent to the claim that $\left(\frac{-2}{p}\right) = 1$ if and only if $p \equiv 1$ or $3 \pmod 8$. To establish the latter, show that $(-2)^{(p-1)/2} \equiv 1 \pmod{p}$ for such primes only and use the hint to Exercise 5.10 below. $\square$

5.3 Dirichlet characters

Fix an integer $m \geq 2$.

A function $\chi\colon \mathbb{Z} \to \mathbb{C}$ is said to be a Dirichlet character modulo m if the following two conditions are satisfied:

(1) $\chi(n) \neq 0$ if and only if $(n, m) = 1$,
(2) χ is periodic, with period m, that is, $\chi(n+m) = \chi(n)$ for all $n \in \mathbb{Z}$, and
(3) χ is completely multiplicative, that is, $\chi(ab) = \chi(a)\chi(b)$ for all $a, b \in \mathbb{Z}$.

The character

$$\chi_0(n) = \begin{cases} 1 & \text{if } (n, m) = 1, \\ 0 & \text{if } (n, m) > 1, \end{cases}$$

will play a special role in our exposition below; it is called the principal character modulo m. Note that $\chi(1) = 1$; this is included in the definition of multiplicative function but is also derivable from $\chi(1^2) = \chi(1)^2$ (condition (1)) and $\chi(1) \neq 0$ (condition (3)).

In order to describe the set of characters modulo m, we will pass to the known algebraic description of the structure of the group $(\mathbb{Z}/m\mathbb{Z})^*$ consisting of residue classes relatively prime with m; in particular, $|(\mathbb{Z}/m\mathbb{Z})^*| = \varphi(m)$. As the group in consideration is commutative (or abelian), the following general result can be applied.

Theorem 5.4. *Every finite abelian group G can be given as a direct product of some of its cyclic subgroups. In other words, there are cyclic subgroups $\langle h_j \rangle_{c_j} = \{h_j^k : k = 0, 1, \dots, c_j - 1\} \subset G$ of order c_j, where $j = 1, \dots, r$, such that*

$$G = \langle h_1 \rangle_{c_1} \cdots \langle h_r \rangle_{c_r} = \{h_1^{k_1} \cdots h_r^{k_r} : 0 \leq k_j < c_j \text{ for } j = 1, \dots, r\}.$$

Then also $|G| = c_1 \cdots c_r$.

Of course, in the case of the group $(\mathbb{Z}/m\mathbb{Z})^*$ there is an explicit description of the direct product decomposition; we will not use it. First, it follows from the Chinese remainder theorem that

$$(\mathbb{Z}/m\mathbb{Z})^* = (\mathbb{Z}/p_1^{\alpha_1}\mathbb{Z})^* \cdots (\mathbb{Z}/p_r^{\alpha_r}\mathbb{Z})^*,$$

where $p_1^{\alpha_1} \cdots p_r^{\alpha_r}$ is the canonical prime factorisation of m. (This is, in fact, a hint to solving Exercise 5.1.) Second, the following exercises show that subgroups $(\mathbb{Z}/p^\alpha\mathbb{Z})^*$ are cyclic for odd primes p, while $(\mathbb{Z}/2^\alpha\mathbb{Z})^*$ is cyclic for $\alpha = 1, 2$ and is a direct product of two cyclic subgroups for $\alpha \geq 3$.

Exercise 5.8. Let p be an odd prime.

(a) Show that the group $(\mathbb{Z}/p\mathbb{Z})^*$ is cyclic, $(\mathbb{Z}/p\mathbb{Z})^* = \langle c\rangle_{p-1}$ for some $c \in \{2, \dots, p-1\}$.

(b) Show that either c or $c+p$ generates the whole group $(\mathbb{Z}/p^\alpha\mathbb{Z})^*$, independent of $\alpha \geq 2$.

Hint. (a) You can use, for example, the fact that $\mathbb{Z}/p\mathbb{Z}$ is a field. □

Exercise 5.9. (a) Verify that $(\mathbb{Z}/2\mathbb{Z})^* = \langle 1\rangle_1$ and $(\mathbb{Z}/4\mathbb{Z})^* = \langle -1\rangle_2$.

(b) Show that, for $\alpha \geq 3$, the group $(\mathbb{Z}/2^\alpha\mathbb{Z})^*$ is the direct product of cyclic subgroups $\langle -1\rangle_2$ and $\langle 5\rangle_{2^{\alpha-2}}$.

Hint. (b) Verify first that 5 has order $c = 2^{\alpha-2}$ in $(\mathbb{Z}/2^\alpha\mathbb{Z})^*$ and use the fact that $\{5^k : k = 0, 1, \dots, c-1\}$ and $\{-5^k : k = 0, 1, \dots, c-1\}$ are disjoint, as they cover different residue classes modulo 4. □

From now on, assume that the group $(\mathbb{Z}/m\mathbb{Z})^*$ is decomposed into a direct product of some of its cyclic subgroups in accordance with Theorem 5.4,

$$(\mathbb{Z}/m\mathbb{Z})^* = \{h_1^{k_1} \cdots h_r^{k_r} \pmod{m} : 0 \leq k_j < c_j \text{ for } j = 1, \dots, r\},$$

where $h_j \pmod{m}$ are generators of the cyclic subgroups of order c_j, where $j = 1, \dots, r$, and $c_1 \cdots c_r = \varphi(m)$. If χ is a character modulo m then $\chi(h_j)^{c_j} = \chi(h_j^{c_j}) = \chi(1) = 1$, hence $\chi(h_j)$ is a root of unity of degree c_j, for each $j = 1, \dots, m$. Suppose we have a collection $\xi_1, \dots, \xi_r$ of roots of unity of respective degrees $c_1, \dots, c_r$. Define the function

$$\chi(a) = \begin{cases} 0 & \text{if } (a, m) \neq 1, \\ \xi_1^{k_1} \cdots \xi_r^{k_r} & \text{if } a \equiv h_1^{k_1} \cdots h_r^{k_r} \pmod{m}. \end{cases}$$

It is not difficult to observe that the properties (1)–(3) in the definition of Dirichlet character are satisfied, thus, χ is a Dirichlet character modulo m. Furthermore, different collections $\xi_1, \dots, \xi_r$ and $\xi'_1, \dots, \xi'_r$ induce different Dirichlet characters χ and χ' modulo m (indeed, since $\xi_j \neq \xi'_j$ for some j, we have $\chi(h_j) \neq \chi'(h_j)$). The correspondence defines a bijection between the group $(\mathbb{Z}/m\mathbb{Z})^*$ and the set of Dirichlet characters modulo m.

Theorem 5.5. *The bijection above is an isomorphism of the groups* $(\mathbb{Z}/m\mathbb{Z})^*$ *and* $\{\chi a \text{ character modulo } m\}$, *where the (commutative) operation in the latter is defined by* $(\chi_1\chi_2)(n) = \chi_1(n)\chi_2(n)$. *In particular, the total number of characters modulo* m *is equal* $\varphi(m)$.

We will heavily use the correspondence between $(\mathbb{Z}/m\mathbb{Z})^*$ and the group of Dirichlet characters modulo m, through collections of roots of unity $\xi_1, \dots, \xi_r$.

Exercise 5.10. For $m = p$ an odd prime, show that the Legendre symbol

$$\chi(a) = \left(\frac{a}{p}\right) = \begin{cases} 0 & \text{if } a \equiv 0 \pmod{p}, \\ 1 & \text{if } x^2 \equiv a \pmod{p} \text{ is soluble in } x \in \mathbb{Z}, \\ -1 & \text{if } x^2 \equiv a \pmod{p} \text{ is not soluble in } x \in \mathbb{Z}, \end{cases}$$

is a Dirichlet character modulo p.

Hint. Use Euler's theorem (Theorem 5.1) to show that

$$a^{(p-1)/2} \equiv \left(\frac{a}{p}\right) \pmod{p}$$

for any $a \in (\mathbb{Z}/p\mathbb{Z})^*$. □

5.4 Properties of Dirichlet characters

Lemma 5.4. *Let $\xi \neq 1$ be a root of unity of degree $c \geq 2$. Then*

$$\sum_{k=0}^{c-1} \xi^k = 0.$$

Proof. Denote $S = \sum_{k=0}^{c-1} \xi^k$. Then $\xi S = \sum_{k=0}^{c-1} \xi^{k+1} = S$ implying $S = 0$. □

Lemma 5.5. *For $0 < k < c$, we have*

$$\sum_{\xi} \xi^k = 0,$$

where the sum is over all roots of unity of degree c.

Proof. Denote $S = \sum_{\xi} \xi^k$ and take a primitive root of unity η of degree c, for example, $\eta = e^{2\pi i/c}$. Then $\eta^k \neq 1$ and it follows from $\eta^k S = \sum_{\xi} (\eta\xi)^k = S$ that $S = 0$. □

Theorem 5.6. *For $m \geq 2$, the value $\chi(n)$ of a character modulo m is a root of unity of degree $\varphi(m)$.*

Furthermore,

$$\sum_{n=1}^{m} \chi(n) = \begin{cases} \varphi(m) & \text{if } \chi = \chi_0 \text{ is the principal character,} \\ 0 & \text{otherwise,} \end{cases}$$

and

$$\sum_{\chi} \chi(n) = \begin{cases} \varphi(m) & \text{if } n \equiv 1 \pmod{m}, \\ 0 & \text{otherwise}, \end{cases}$$

where the latter summation is over all characters modulo m.

Proof. The first part follows from the isomorphism in Theorem 5.5. Indeed, we have $\chi(n) = \xi_1^{k_1} \cdots \xi_r^{k_r}$ for n relatively prime to m, where $\xi_1, \ldots, \xi_r$ are roots of unity of respective degrees $c_1, \ldots, c_r$. But then $\chi(n)^{c_1 \cdots c_r} = 1$ where $c_1 \cdots c_r = \varphi(m)$.

Notice that

$$\sum_{n=1}^{m} \chi(n) = \sum_{\substack{n=1 \\ (n,m)=1}}^{m} \chi(n)$$

and the latter sum consists of $\varphi(m)$ terms equal to 1 if χ is the principal character. If χ is not, then the corresponding collection $\xi_1, \ldots, \xi_r$ contains at least one root of unity different from 1, so that the sum

$$\sum_{\substack{n=1 \\ (n,m)=1}}^{m} \chi(n) = \sum_{\substack{0 \le k_j < c_j \\ j=1,\ldots,r}} \xi_1^{k_1} \xi_2^{k_2} \cdots \xi_r^{k_r} = \sum_{k_1=0}^{c_1-1} \xi_1^{k_1} \times \sum_{k_2=0}^{c_2-1} \xi_2^{k_2} \times \cdots \times \sum_{k_r=0}^{c_r-1} \xi_r^{k_r}$$

vanishes as at least one of its factors does (by Lemma 5.4).

Similarly, in the case of the sum over characters, the condition $n \not\equiv 1 \pmod{m}$, $(n,m) = 1$, means that for at least one character we get $\chi(n) = \xi_1^{k_1} \cdots \xi_r^{k_r} \neq 1$, so that at least one exponent k_j is strictly between 0 and c_j. Then

$$\sum_{\chi} \chi(n) = \sum_{\xi_1, \xi_2, \ldots, \xi_r} \xi_1^{k_1} \xi_2^{k_2} \cdots \xi_r^{k_r} = \sum_{\xi_1} \xi_1^{k_1} \times \sum_{\xi_2} \xi_2^{k_2} \times \cdots \times \sum_{\xi_r} \xi_r^{k_r} = 0,$$

since the factor $\sum_{\xi_j} \xi_j^{k_j}$ vanishes by Lemma 5.5. □

Lemma 5.6. *For real $x \ge 1$ and a non-principal character χ modulo m, the sum*

$$S(x) = \sum_{1 \le n \le x} \chi(n)$$

is bounded: $|S(x)| \le m$.

Proof. As the character χ is a periodic function of period m and the sum $\sum_{n=1}^{m}\chi(n)$ vanishes on the full period, we only need to check the inequality for $1 \le x < m$. In this case,

$$|S(x)| \le \sum_{1\le n\le x} |\chi(n)| < \sum_{1\le n\le m} 1 < m. \qquad \square$$

Lemma 5.7. *To an integer a, $(a,m)=1$, assign the least positive exponent f for which $a^f \equiv 1 \pmod{m}$. Then the set $\{\chi(a) : \chi$ is a character modulo $m\}$, in which numbers appear multiple times, is the set of roots of unity of degree f such that each root of unity appears exactly $\varphi(m)/f$ times.*

Proof. The fact that all entries in $\{\chi(a) : \chi\}$ are roots of unity of degree f is straightforward: $\chi(a)^f = \chi(a^f) = \chi(1) = 1$. Now take a root of unity ξ of degree f. From the last formula in Theorem 5.6,

$$\begin{aligned} S &= \sum_{\chi}(\xi^{-1}\chi(a) + \xi^{-2}\chi(a^2) + \cdots + \xi^{-f}\chi(a^f)) \\ &= \sum_{k=1}^{f}\xi^{-k}\sum_{\chi}\chi(a^k) = \xi^{-f}\sum_{\chi}\chi(a^f) = \varphi(m). \end{aligned}$$

On the other hand, the same sum can be computed using Lemma 5.4 and the fact that $\xi^{-1}\chi(a)$ is a root of unity of degree f (as both ξ and $\chi(a)$ are). We have

$$\sum_{k=1}^{f}\xi^{-k}\chi(a^k) = \sum_{k=1}^{f}(\xi^{-1}\chi(a))^k = \begin{cases} 0 & \text{if } \xi \ne \chi(a), \\ f & \text{if } \xi = \chi(a). \end{cases}$$

Therefore, if q is the number of characters χ for which $\chi(a) = \xi$, then $S = qf$. Combining this with the computation of S above we deduce that $q = \varphi(m)/f$. $\square$

5.5 Dirichlet L-functions and their basic properties

For χ a character modulo $m \ge 2$, the Dirichlet L-function is defined by the series

$$L(s,\chi) = \sum_{n=1}^{\infty}\frac{\chi(n)}{n^s}.$$

As in Chapter 2 we will write $s = \sigma + it$.

Lemma 5.8. *The series defining the Dirichlet L-function converges in the half-plane* $\operatorname{Re} s > 0$ *for non-principal characters; the convergence is in the half-plane* $\operatorname{Re} s > 1$ *for the principal character. The function* $L(s, \chi)$ *is analytic in the corresponding domain, and its consequent derivatives can be computed by term-wise differentiation of the series.*

Proof. If $\chi = \chi_0$ is the principal character and $\delta > 0$ is arbitrary, then

$$\left| \frac{\chi(n)}{n^s} \right| \le \frac{1}{n^\sigma} \le \frac{1}{n^{1+\delta}}$$

for $\operatorname{Re} s > 1 + \delta$ implying that the sums

$$\left| \sum_{n=1}^{\infty} \frac{\chi(n)}{n^s} \right| \le \sum_{n=1}^{\infty} \frac{1}{n^{1+\delta}} = \text{const}$$

are uniformly bounded in the domain. Thus, the series for $L(s, \chi_0)$ converges uniformly there and define an analytic function by the Weierstrass theorem. Since $\delta > 0$ is arbitrary, this function $L(s, \chi_0)$ is analytic in the domain $\operatorname{Re} s > 1$.

Assume now that the character χ is non-principal and define the sum $S(x) = \sum_{1 \le n \le x} \chi(n)$, so that $\chi(n) = S(n) - S(n-1)$ and

$$\begin{aligned}\sum_{n=1}^{N} \frac{\chi(n)}{n^s} &= \sum_{n=1}^{N} \frac{S(n) - S(n-1)}{n^s} = \sum_{n=1}^{N} \frac{S(n)}{n^s} - \sum_{n=1}^{N-1} \frac{S(n)}{(n+1)^s} \\ &= \sum_{n=1}^{N} S(n) \left(\frac{1}{n^s} - \frac{1}{(n+1)^s} \right) + \frac{S(N)}{(N+1)^s}.\end{aligned}$$

By Lemma 5.6, $|S(n)| \le m$; in particular, this implies that $S(N)/(N+1)^s \to 0$ as $N \to \infty$. For the terms of the sum we have

$$\begin{aligned}\left| S(n) \left(\frac{1}{n^s} - \frac{1}{(n+1)^s} \right) \right| &= \left| S(n) s \int_n^{n+1} t^{-s-1} \mathrm{d}t \right| \le m|s| \int_n^{n+1} t^{-\sigma-1} \mathrm{d}t \\ &\le \frac{m|s|}{n^{1+\sigma}} \le \frac{m|s|}{n^{1+\delta}}\end{aligned}$$

in the domain $\operatorname{Re} s > \delta$. Since the dominant series

$$|s| \sum_{n=1} \frac{1}{n^{1+\delta}}$$

converges, we conclude, again appealing to the Weierstrass theorem, that the sequence of partial sums

$$\sum_{n=1}^{N} \frac{\chi(n)}{n^s}$$

uniformly converges to an analytic function in the domain $\operatorname{Re} s > \delta$, $|s| \le M$; thus, $L(s,\chi)$ is analytic in the half-plane $\operatorname{Re} s > 0$ and we can differentiate its series representation there term-wise. □

Lemma 5.9. *In the half-plane* $\operatorname{Re} s > 1$, *the representation*

$$-\frac{L'(s,\chi)}{L(s,\chi)} = \sum_{n=1}^{\infty} \frac{\chi(n)\Lambda(n)}{n^s}$$

is valid, where $\Lambda(n)$ *is the von Mangoldt function (see Section* 2.2*). In particular,* $L(s,\chi)$ *does not vanish in the half-plane.*

Proof. The proof of this statement is exactly the same as of Lemma 2.4 (and Theorem 2.2) earlier. From Lemma 5.8 we have

$$L'(s,\chi) = -\sum_{n=1}^{\infty} \frac{\chi(n)\ln n}{n^s}$$

in the half-plane $\operatorname{Re} s > 1$, while

$$\begin{aligned}\sum_{n=1}^{\infty} \frac{\chi(n)\ln n}{n^s} &= \sum_{n=1}^{\infty} \frac{\chi(n)}{n^s} \sum_{k|n} \Lambda(k) = \sum_{l=1}^{\infty}\sum_{k=1}^{\infty} \frac{\chi(lk)\Lambda(k)}{(lk)^s} \\ &= \sum_{l=1}^{\infty} \frac{\chi(l)}{l^s} \times \sum_{k=1}^{\infty} \frac{\chi(k)\Lambda(k)}{k^s} = L(s,\chi)\sum_{k=1}^{\infty} \frac{\chi(k)\Lambda(k)}{k^s},\end{aligned}$$

from which the required identity follows. The vanishing of $L(s,\chi)$ for some s with $\operatorname{Re} s > 1$ would produce a pole of the logarithmic derivative $L'(s,\chi)/L(s,\chi)$, while the series representation guarantees none. □

5.6 Euler's product for Dirichlet L-functions. Analytic continuation to the domain $\operatorname{Re} s > 0$

Applying Lemma 2.6 with the choice $f(n) = \chi(n)/n^s$, where χ is a character modulo m, we arrive at the following Euler-type product identity for $L(s,\chi)$.

Theorem 5.7. *In the half-plane* $\operatorname{Re} s > 1$, *we have*

$$L(s,\chi) = \prod_p \left(1 - \frac{\chi(p)}{p^s}\right)^{-1}$$

where the product is over all primes.

One important corollary of this identity is a simple recipe to continue $L(s,\chi_0)$ analytically to the half-plane $\operatorname{Re} s > 0$.

Lemma 5.10. *The formula*

$$L(s, \chi_0) = \zeta(s) \prod_{p|m} \left(1 - \frac{1}{p^s}\right)$$

defines the analytic continuation of $L(s, \chi_0)$ *to the domain* $\operatorname{Re} s > 0$. *It has a single singular point there — the pole of order* 1 *at* $s = 1$, *with residue* $\prod_{p|m}(1 - 1/p) \neq 0$.

Proof. It follows from Theorems 5.7 and 2.4 that

$$L(s, \chi_0) = \prod_{p} \left(1 - \frac{\chi_0(p)}{p^s}\right)^{-1} = \prod_{p\nmid m} \left(1 - \frac{\chi_0(p)}{p^s}\right)^{-1} = \zeta(s) \prod_{p|m} \left(1 - \frac{1}{p^s}\right)$$

in the half-plane $\operatorname{Re} s > 1$. As $\zeta(s)$ is an analytic function in the larger domain $\operatorname{Re} s > 0$, and its only singular point there is the single simple pole with residue 1 at $s = 1$, we deduce the related implications for the function $L(s, \chi_0)$ as well. □

5.7 The nonvanishing of $L(1, \chi)$ for non-principal characters χ

Given $m \geq 2$, we now turn to studying the properties of the product

$$F(s) = \prod_{\chi} L(s, \chi),$$

where the product is over all characters modulo m. As each of the multiples in the product is a series of the form

$$\sum_{n=1}^{\infty} \frac{b_n}{n^s}$$

for some $b_n \in \mathbb{C}$, the product also has this form.

Lemma 5.11. *The expansion*

$$F(s) = \sum_{n=1}^{\infty} \frac{a_n}{n^s}$$

represents an analytic function in the domain $\operatorname{Re} s > 1$, *for which the derivatives can be computed by term-wise differentiation:*

$$F^{(k)}(s) = (-1)^k \sum_{n=1}^{\infty} \frac{a_n (\ln n)^k}{n^s}, \quad \text{where } k = 1, 2, \ldots .$$

Furthermore, the coefficients a_n *are non-negative integers, and if* $n = q^{\varphi(m)}$ *for some* q *coprime with* m, *then* a_n *are positive integers.*

Proof. Fix a prime p such that $p \nmid m$ and denote by f the least positive exponent for which $p^f \equiv 1 \pmod{m}$. By Lemma 5.7 the set $\{\chi(p) : \chi\}$ consists of all roots of unity of degree f, each occurred exactly $g = \varphi(m)/f$ times. Denote $\xi_1, \dots, \xi_f$ all such roots of unity, so that $(1 - \xi_1 t) \cdots \times (1 - \xi_f t) = 1 - t^f$. Taking $t = 1/p^s$ and using the binomial formula

$$(1-x)^{-g} = \sum_{r=0}^{\infty} \frac{(-g)(-g-1)\cdots(-g-r+1)}{r!}(-x)^r = \sum_{r=0}^{\infty} \binom{g+r-1}{g-1} x^r$$

when $|x| < 1$, we therefore obtain

$$\begin{aligned}\prod_{\chi}\left(1 - \frac{\chi(p)}{p^s}\right)^{-1} &= \left(1 - \left(\frac{1}{p^s}\right)^f\right)^{-g} = \left(1 - \frac{1}{p^{fs}}\right)^{-g} \\ &= \sum_{r=0}^{\infty} \binom{g+r-1}{g-1} p^{-frs} = \sum_{k=0}^{\infty} u_{p,k} p^{-ks}\end{aligned}$$

where

$$u_{p,k} = \begin{cases} 0 & \text{if } f \nmid k, \\ \dbinom{g+k/f-1}{g-1} & \text{if } f \mid k. \end{cases}$$

It follows from the latter expansion that

$$\begin{aligned}\prod_{\chi}\prod_{p \le N}\left(1 - \frac{\chi(p)}{p^s}\right)^{-1} &= \prod_{\substack{p \le N \\ p \nmid m}}\prod_{\chi}\left(1 - \frac{\chi(p)}{p^s}\right)^{-1} = \prod_{\substack{p \le N \\ p \nmid m}}\left(\sum_{k=0}^{\infty} u_{p,k} p^{-ks}\right) \\ &= \sum_{\substack{(p_1,\dots,p_l):p_j \le N \\ k_1,\dots,k_l \ge 0}} u_{p_1,k_1} \cdots u_{p_l,k_l} p_1^{-k_1 s} \cdots p_l^{-k_l s} \\ &= \sum_{n=1}^{\infty}{}' \frac{a_n}{n^s},\end{aligned}$$

where the integers n involved in the latter sum $\sum'$ all have their prime divisors at most N (and relatively prime with m); here

$$a_n = \begin{cases} u_{p_1,k_1} \cdots u_{p_l,k_l} & \text{if } (n,m) = 1 \text{ and } n = p_1^{k_1} \cdots p_l^{k_l}, \\ 0 & \text{if } (n,m) \ne 1. \end{cases}$$

This expression means that a_n are always non-negative integers; if $n = q^{\varphi(m)}$ then all exponents in the prime factorisation of n are divisible by $\varphi(m)$, so that all u_{p_j,k_j} are positive integers.

Take now an arbitrary $\sigma > 1$. Then

$$\sum_{n=1}^{N} \frac{a_n}{n^\sigma} \le \sum_{n=1}^{\infty}{}' \frac{a_n}{n^\sigma} = \prod_{\chi} \prod_{p \le N} \left(1 - \frac{\chi(p)}{p^\sigma}\right)^{-1}$$

and

$$\left| \prod_{\chi} \prod_{p \le N} \left(1 - \frac{\chi(p)}{p^\sigma}\right)^{-1} \right| \le \left| \prod_{\chi} \prod_{p} \left(1 - \frac{\chi(p)}{p^\sigma}\right)^{-1} \right| = |F(\sigma)|,$$

so that all the partial sums of the series

$$\sum_{n=1}^{N} \frac{a_n}{n^\sigma}$$

of non-negative terms are bounded above by the constant $|F(\sigma)|$ independent of N. We wish to demonstrate the uniform convergence of

$$\sum_{n=1}^{N} \frac{a_n}{n^s}$$

to $F(s)$ in the domain $\operatorname{Re} s \ge \sigma$. We have

$$\begin{aligned} \left| F(s) - \sum_{n=1}^{N} \frac{a_n}{n^s} \right| &\le \left| F(s) - \prod_{\chi} \prod_{p \le N} \left(1 - \frac{\chi(p)}{p^s}\right)^{-1} \right| \\ &\quad + \left| \prod_{\chi} \prod_{p \le N} \left(1 - \frac{\chi(p)}{p^s}\right)^{-1} - \sum_{n=1}^{N} \frac{a_n}{n^s} \right| \\ &< \frac{\varepsilon}{2} + \left| \sum_{n=1}^{\infty}{}' \frac{a_n}{n^s} - \sum_{n=1}^{N} \frac{a_n}{n^s} \right| \\ &\le \frac{\varepsilon}{2} + \sum_{n \ge N+1}{}' \left| \frac{a_n}{n^s} \right| = \frac{\varepsilon}{2} + \sum_{n \ge N+1}{}' \frac{a_n}{n^\sigma} < \varepsilon \end{aligned}$$

as $N \to \infty$, with all the estimates uniform in the domain. Then the Weierstrass theorem guarantees the uniform convergence and term-wise differentiation for $\operatorname{Re} s > \sigma$, hence for $\operatorname{Re} s > 1$, since $\sigma > 1$ is chosen arbitrary. □

Lemma 5.12. *If $F(s)$ is analytic in the domain* $\operatorname{Re} s > 0$, *then the series*

$$\sum_{n=1}^{\infty} \frac{a_n}{n^s}$$

constructed in Lemma 5.11 *converges to $F(s)$ in the domain.*

Proof. Assuming that $F(s)$ is analytic in the half-plain $\operatorname{Re} s > 0$, expand the function into its Taylor series at $s = 2 + it_0$:

$$F(s) = \sum_{k=0}^{\infty} \frac{F^{(k)}(2+it_0)}{k!} (s - (2+it_0))^k.$$

The series converges in the open disc $|s - (2+it_0)| < 2$ of radius 2, as it entirely lies in the domain of analyticity of $F(s)$. Then for $0 < \sigma \leq 2$, we have

$$\begin{aligned}
F(\sigma + it_0) &= \sum_{k=0}^{\infty} \frac{F^{(k)}(2+it_0)}{k!} (\sigma - 2)^k = \sum_{k=0}^{\infty} \frac{(\sigma-2)^k}{k!} (-1)^k \sum_{n=1}^{\infty} \frac{a_n (\ln n)^k}{n^{2+it_0}} \\
&= \sum_{k=0}^{\infty} \sum_{n=1}^{\infty} \frac{a_n (2-\sigma)^k (\ln n)^k}{k!\, n^{2+it_0}} = \sum_{n=1}^{\infty} \sum_{k=0}^{\infty} \frac{a_n (2-\sigma)^k (\ln n)^k}{k!\, n^{2+it_0}} \\
&= \sum_{n=1}^{\infty} \frac{a_n}{n^{2+it_0}} \sum_{k=0}^{\infty} \frac{(2-\sigma)^k (\ln n)^k}{k!} = \sum_{n=1}^{\infty} \frac{a_n}{n^{2+it_0}} e^{(2-\sigma)\ln n} \\
&= \sum_{n=1}^{\infty} \frac{a_n}{n^{2+it_0}} \times n^{2-\sigma} = \sum_{n=1}^{\infty} \frac{a_n}{n^{\sigma+it_0}},
\end{aligned}$$

which establishes the validity of the series expansion for $F(s)$ for any s with $\operatorname{Re} s > 0$, as the choice of t_0 is arbitrary. It remains to show that the interchange of summation over n and over k is legitimate. Indeed, sums of the terms

$$\frac{a_n (2-\sigma)^k (\ln n)^k}{k!\, n^{2+it_0}}$$

converge uniformly in $\operatorname{Re} s \geq \delta$ for any choice of $\delta > 0$, in either n or k. This is true because all a_n are non-negative real numbers, so that

$$\left| \sum_{n=1}^{\infty} \frac{a_n (2-\sigma)^k (\ln n)^k}{k!\, n^{2+it_0}} \right| \leq \frac{(2-\sigma)^k}{k!} \sum_{n=1}^{\infty} \frac{a_n (\ln k)^n}{n^2} = \frac{(2-\sigma)^k}{k!} F^{(k)}(2)$$

and

$$\left| \sum_{k=0}^{\infty} \frac{a_n (2-\sigma)^k (\ln n)^k}{k!\, n^{2+it_0}} \right| = \frac{a_n}{n^2} \sum_{k=0}^{\infty} \frac{(2-\sigma)^k (\ln n)^k}{k!} = \frac{a_n}{n^2} e^{(2-\sigma)\ln n}$$

for $\operatorname{Re} s \geq \delta$. □

Lemma 5.13. *$L(1, \chi) \neq 0$ for any non-principal character χ modulo m.*

Proof. Assume on the contrary that is false and $L(1,\chi)=0$ for at least one character χ modulo m. Then $F(s)$ is analytic in the half-plane $\operatorname{Re} s>0$, since the only potential pole of $F(s)$ contributed by $L(s,\chi_0)$ at $s=1$ (according to Lemma 5.10) is cancelled by the zero of $L(s,\chi)$ at $s=1$. By Lemma 5.12, the expansion

$$F(s)=\sum_{n=1}^{\infty}\frac{a_n}{n^s}$$

is valid for $F(s)$ in the domain $\operatorname{Re} s>0$, so that the series on the right-hand side converges for any such s. On the other hand, $a_n\ge 0$ and $a_n\ge 1$ for any n of the form $n=r^{\varphi(m)}$, where $(r,m)=1$. By choosing $s_0=1/\varphi(m)$ we obtain a presumably convergent series,

$$F\Big(\frac{1}{\varphi(m)}\Big)=\sum_{n=1}^{\infty}\frac{a_n}{n^{1/\varphi(m)}}\ge\sum_{\substack{k=0\\ n=(mk+1)^{\varphi(m)}}}^{\infty}\frac{1}{n^{1/\varphi(m)}}=\sum_{k=0}^{\infty}\frac{1}{mk+1}=\infty;$$

which is a contradiction. Thus, $F(s)$ cannot be analytic in the domain $\operatorname{Re} s>0$ meaning that none of $L(s,\chi)$ vanishes at $s=1$. □

Exercise 5.11. For a character χ modulo m, consider the generating function

$$Q(x,\chi)=\sum_{n=0}^{\infty}\chi(n)x^n.$$

(a) Show that series for $Q(x,\chi)$ converges in the disc $|x|<1$ and that 1 is its radius of convergence.
(b) Prove that $Q(x,\chi)$ is a rational function of x.
(c) For χ non-principal character, show that

$$L(1,\chi)=\int_0^1 Q(x,\chi)\,\frac{\mathrm{d}x}{x}.$$

(d) For non-principal characters χ modulo 3 and modulo 4, compute the quantities $L(1,\chi)$ and check they are nonzero.

5.8 Proof of Dirichlet's theorem on primes in arithmetic progressions

Theorem 5.8 (Dirichlet's theorem). *Let $m\ge 2$ and l be integers, $(l,m)=1$. Then the arithmetic progression $p\equiv l \pmod m$ contains infinitely many primes.*

Proof. For a fixed character χ modulo m, consider the logarithmic derivative of $L(s,\chi)$ in the half-plane $\sigma = \operatorname{Re} s > 1$, where the convergence of the series is absolute:

$$-\frac{L'(s,\chi)}{L(s,\chi)} = \sum_{n=1}^{\infty} \frac{\chi(n)\Lambda(n)}{n^s} = \sum_p \sum_{k=1}^{\infty} \frac{\chi(p^k)\ln p}{p^{ks}} = \sum_p \ln p \sum_{k=1}^{\infty} \left(\frac{\chi(p)}{p^s}\right)^k$$
$$= \sum_p \ln p \times \frac{\chi(p)/p^s}{1-\chi(p)/p^s} = \sum_p \ln p \left(\frac{\chi(p)}{p^s} + \frac{\chi(p)^2}{p^{2s}(1-\chi(p)/p^s)}\right)$$
$$= \sum_p \frac{\chi(p)\ln p}{p^s} + \sum_p \frac{\chi(p)^2 \ln p}{p^s(p^s-\chi(p))}.$$

In the second sum we have $|p^s - \chi(p)| \ge |p^s| - 1 = p^\sigma - 1 > \frac{1}{2}p^\sigma$, because $p^\sigma > 2$ for $\sigma > 1$; therefore,

$$\left|\sum_p \frac{\chi(p)^2 \ln p}{p^s(p^s-\chi(p))}\right| \le \sum_p \frac{|\chi(p)^2| \ln p}{|p^s|\,|p^s-\chi(p)|}$$
$$< \sum_p \frac{2\ln p}{p^{2\sigma}} < \sum_p \frac{2\ln p}{p^2} < \sum_{n=2}^{\infty} \frac{2\ln n}{n^2}$$

is the uniform bound for the sum in the domain $\operatorname{Re} s > 1$. We conclude that

$$\sum_p \frac{\chi(p)}{p^s} \ln p = -\frac{L'(s,\chi)}{L(s,\chi)} + O(1) \quad \text{as } s \to 1^+$$

along the real axis.

Define an integer v such that $vl \equiv 1 \pmod{m}$; it represents a particular residue class in $(\mathbb{Z}/m\mathbb{Z})^*$. It follows from Theorem 5.6 that

$$\sum_p \frac{\ln p}{p^s} \sum_\chi \chi(pv) = \sum_{p:pv\equiv 1 \pmod{m}} \frac{\ln p}{p^s}\varphi(m) = \varphi(m) \sum_{p:p\equiv l \pmod{m}} \frac{\ln p}{p^s},$$

while from the asymptotics above we obtain

$$\sum_p \frac{\ln p}{p^s} \sum_\chi \chi(pv) = \sum_\chi \chi(v) \sum_p \frac{\ln p}{p^s}\chi(p) = -\sum_\chi \chi(v)\frac{L'(s,\chi)}{L(s,\chi)} + O(1)$$
$$= -\chi_0(v)\frac{L'(s,\chi_0)}{L(s,\chi_0)} - \sum_{\chi\ne\chi_0} \chi(v)\frac{L'(s,\chi)}{L(s,\chi)} + O(1)$$
$$= -\frac{L'(s,\chi_0)}{L(s,\chi_0)} + O(1)$$

as $s \to 1^+$, because from Lemma 5.13 the functions $L'(s,\chi)/L(s,\chi)$ are analytic at $s = 1$ for all non-principal characters χ. We know from

Lemma 5.10 that the function $L(s,\chi_0)$ has a simple pole at $s=1$; hence $-L'(s,\chi_0)/L(s,\chi_0)$ has pole of order 1 there with residue 1. Thus,

$$\varphi(m)\sum_{p:p\equiv l \pmod m}\frac{\ln p}{p^s}=\frac{1}{s-1}+O(1)$$

as $s\to 1^+$ meaning that the series

$$\sum_{p:p\equiv l \pmod m}\frac{\ln p}{p}$$

diverges; in particular, it involves infinitely many terms. □

Chapter notes

The first proof of Wilson's theorem (1770), which is stated in Exercise 5.3, was given by Lagrange in 1771.

The proof of Dirichlet's theorem given by Dirichlet (1837) introduces the notions of Dirichlet characters and Dirichlet L-functions; these inspired several branches in mathematics to appear, with interconnections between those unified by Langlands' programme. Particular instances of the latter are exemplified [75] through the quadratic Legendre(–Jacobi–Kronecker) symbols and the reciprocity law for them, and this study traces back to Euler.

Though we do not pursue quantitative aspects of Dirichlet's theorem on primes in arithmetic progressions, one can naturally extend the argument of Chapter 2 to show that

$$\sum_{\substack{p\le x\\ p\equiv l \pmod m}}1\sim\frac{1}{\varphi(m)}\frac{x}{\ln x}\qquad\text{as } x\to\infty,$$

when $(l,m)=1$. The question about representativeness of primes as values of a single-variable polynomial of degree at least 2 (for example, of u^2+1) is pretty open. Proven results for two-variable polynomials (then of degree at least 3, to distinguish from 'easier' situations like in Theorem 4.1) — there are infinitely many prime numbers of the form u^2+v^4 by J. Friedlander and H. Iwaniec [32] and of the form u^3+2v^3 by R. Heath-Brown [49] — are already at the boundary of possible for the current methods.

Chapter 6

Algebraic and transcendental numbers. The transcendence of e and π

6.1 Algebraic numbers: basic properties and algebraic closedness

A complex number α is said to be *algebraic* if there is a polynomial $f(x) \not\equiv 0$ with rational coefficients such that $f(\alpha) = 0$. Rational numbers are basic examples of algebraic numbers, when such polynomials can be taken linear.

Lemma 6.1. *Let α be algebraic and $f(x) \in \mathbb{Q}[x]$ a non-trivial polynomial of least possible degree such that $f(\alpha) = 0$. Then $f(x)$ is irreducible over $\mathbb{Q}$. Furthermore, if $g(x) \in \mathbb{Q}[x]$ and $g(\alpha) = 0$ then $f(x)$ divides $g(x)$.*

Proof. Assume that $f(x)$ is reducible over $\mathbb{Q}$, that is, $f(x) = u(x)v(x)$ for some polynomials $u(x)$ and $v(x)$ with rational coefficients, of *smaller* degrees. Then $0 = f(\alpha) = u(\alpha)v(\alpha)$ implying $u(\alpha) = 0$ or $v(\alpha) = 0$, thus, contradicting to the minimality of degree hypothesis on f.

For the second part, divide $g(x)$ by $f(x)$ in $\mathbb{Q}[x]$ with remainder: $g(x) = q(x)f(x) + r(x)$, where $r(x) \equiv 0$ or $\deg r(x) < \deg f(x)$. Then

$$0 = g(\alpha) = q(\alpha)f(\alpha) + r(\alpha) = r(\alpha),$$

hence the minimality of degree hypothesis on f implies $r(x) \equiv 0$. □

An irreducible polynomial from Lemma 6.1 is called a *minimal polynomial* of α. It follows from the lemma all such minimal polynomials of a given algebraic number α are rationally proportional to each other.

Exercise 6.1. Show that for each positive integer n, the polynomial $x^n - 2$ is irreducible.

Exercise 6.2 (Eisenstein's criterion). Assume that for a polynomial

$$f(x) = x^n + a_{n-1}x^{n-1} + \cdots + a_1x + a_0 \in \mathbb{Z}[x]$$

there is a prime p dividing all the coefficients $a_{n-1}, \dots, a_1, a_0$ but $p^2 \nmid a_0$. Show that $f(x)$ is irreducible.

Exercise 6.3. Show that the cyclotomic polynomial $\Phi_p(x) = x^{p-1} + x^{p-2} + \cdots + x + 1$ is irreducible.

Hint. Use Eisenstein's criterion. □

Lemma 6.2. *Suppose that a polynomial $f(x) \in \mathbb{Q}[x]$ is irreducible and has a common zero with a polynomial $g(x) \in \mathbb{Q}[x]$. Then $f(x)$ divides $g(x)$, and all zeros of $f(x)$ are zeros of $g(x)$ as well.*

Proof. Let α be the common zero of $f(x)$ and $g(x)$; it is algebraic. From the irreducibility of $f(x)$ and Lemma 6.1, the polynomial is a minimal polynomial for α and it divides $g(x)$. This immediately implies the claim. □

The *degree* of an algebraic number α is the degree of its minimal polynomial. In particular, rational numbers have degree 1, while quadratic irrationalities, like $i = \sqrt{-1}$, have degree 2.

Lemma 6.3. *Suppose that $\gamma_1, \dots, \gamma_m$ are complex numbers, not all zero; define their span*

$$L = \{r_1\gamma_1 + \cdots + r_m\gamma_m : r_j \in \mathbb{Q} \text{ for } j = 1, \dots, m\}$$

over $\mathbb{Q}$ (also known as a $\mathbb{Q}$-lattice). If $\lambda L \subset L$ then λ is algebraic of degree at most m.

Proof. Indeed, the inclusion is equivalent to the system of linear equations

$$\begin{aligned}
\lambda\gamma_1 &= r_{11}\gamma_1 + r_{12}\gamma_2 + \cdots + r_{1m}\gamma_m, \\
\lambda\gamma_2 &= r_{21}\gamma_1 + r_{22}\gamma_2 + \cdots + r_{2m}\gamma_m, \\
&\cdots\cdots\cdots\cdots\cdots\cdots \\
\lambda\gamma_m &= r_{m1}\gamma_1 + r_{m2}\gamma_2 + \cdots + r_{mm}\gamma_m,
\end{aligned}$$

or

$$\begin{aligned}
0 &= (r_{11} - \lambda)\gamma_1 + r_{12}\gamma_2 + \cdots + r_{1m}\gamma_m, \\
0 &= r_{21}\gamma_1 + (r_{22} - \lambda)\gamma_2 + \cdots + r_{2m}\gamma_m, \\
&\cdots\cdots\cdots\cdots\cdots\cdots \\
0 &= r_{m1}\gamma_1 + r_{m2}\gamma_2 + \cdots + (r_{mm} - \lambda)\gamma_m.
\end{aligned}$$

The determinant of the associated matrix, $f(x) = \det_{1 \le j,k \le m}(r_{jk} - x\delta_{jk})$, vanishes at $x = \lambda$, as the system has a nonzero solution $(\gamma_1, \dots, \gamma_m)$. Since

$f(x)$ is a polynomial with rational coefficients of degree at most m, the desired claim follows. □

Theorem 6.1. *Let α and β be two algebraic numbers. Then $\alpha+\beta$, $\alpha-\beta$, $\alpha\beta$ and α/β (if $\beta \neq 0$) are algebraic, each of degree not more than the product of degrees of α and of β.*

Proof. Denote $f(x) = x^n + a_{n-1}x^{n-1} + \cdots + a_0$ a (monic) minimal polynomial of α, of degree n, and $g(x)$ a minimal polynomial of β, of degree k. Set $m = nk$ and $\{\gamma_j : j = 1, \dots, m\} = \{\alpha^r\beta^s : r = 0, 1, \dots, n-1;\ s = 0, 1, \dots, k-1\}$, and define the rational span L of the latter set as in Lemma 6.3. Then $\alpha L \subset L$. Indeed, $\alpha \cdot \alpha^r\beta^s = \alpha^{r+1}\beta^s$ is one of the elements γ_j if $r < n-1$, while

$$\alpha^{r+1}\beta^s = \alpha^n\beta^s = -a_0\beta^s - \cdots - a_{n-1}\alpha^{n-1}\beta^s \in L$$

if $r = n-1$. For the same reason, $\beta L \subset L$. The inclusions imply

$$(\alpha \pm \beta)L = \alpha L \pm \beta L \subset L \quad \text{and} \quad (\alpha\beta)L = \alpha(\beta L) \subset \alpha L \subset L,$$

so that the numbers $\alpha \pm \beta$ and $\alpha\beta$ are algebraic of degree at most m by Lemma 6.3. For the last part, observe that for $\beta \neq 0$ its reciprocal β^{-1} is a zero of the polynomial $x^k g(1/x)$ of degree k, hence it is algebraic of degree at most k. From the above, we conclude that $\alpha/\beta = \alpha \cdot \beta^{-1}$ is algebraic of degree at most $m = nk$. □

Exercise 6.4. Show that if $\beta \neq 0$ is an algebraic number of degree k then its reciprocal $1/\beta$ is also algebraic of the same degree.

Theorem 6.2 (algebraic closedness of algebraic numbers). *Let $f(x) = x^n + \alpha_{n-1}x^{n-1} + \cdots + \alpha_1 x + \alpha_0$ be a polynomial with algebraic coefficients $\alpha_{n-1}, \dots, \alpha_1, \alpha_0$ and $\alpha \in \mathbb{C}$ a zero of $f(x)$. Then α is algebraic.*

Proof. For each algebraic coefficient α_j of the polynomial, denote by k_j its degree. This time, the set L in Lemma 6.3 is the $\mathbb{Q}$-span of the following numerical collection:

$$\{\alpha^j\alpha_0^{j_0}\alpha_1^{j_1}\cdots\alpha_{n-1}^{j_{n-1}} : 0 \le j < n;\ 0 \le j_0 < k_0;\\ 0 \le j_1 < k_1;\ \dots;\ 0 \le j_{n-1} < k_{n-1}\}.$$

As in the proof of Theorem 6.1 we clearly have $\alpha_j L \subset L$ for $j = 0, 1, \dots, n-1$. In a similar way, $\alpha L \subset L$, because

$$\alpha \cdot \alpha^j\alpha_0^{j_0}\cdots\alpha_{n-1}^{j_{n-1}} = \alpha^{j+1}\alpha_0^{j_0}\cdots\alpha_{n-1}^{j_{n-1}}$$

which is a number from the collection for $j < n-1$ and is equal to

$$(-\alpha_0 - \alpha_1\alpha - \cdots - \alpha_{n-1}\alpha^{n-1})\alpha_0^{j_0}\cdots\alpha_{n-1}^{j_{n-1}} \in L$$

for $j = n-1$. It follows from Lemma 6.3 that α is algebraic of degree at most $nk_0k_1\cdots k_{n-1}$. □

6.2 Rational approximations of real numbers. Non-quadraticity of e

Theorem 6.3. *Let α be a real number and t a positive integer. Then there is a rational number p/q such that*

$$\left|\alpha - \frac{p}{q}\right| < \frac{1}{qt}$$

and $1 \le q \le t$.

Proof. The proof makes use of an elementary argument known as Dirichlet's box principle or as the pigeon hole principle. For the fractional parts,

$$\{\alpha x\} = \alpha x - \lfloor \alpha x \rfloor \in [0,1) = \left[0, \frac{1}{t}\right) \cup \left[\frac{1}{t}, \frac{2}{t}\right) \cup \cdots \cup \left[\frac{t-1}{t}, 1\right),$$

when x runs over integers $0, 1, \dots, t$, at least two of the values of $\{\alpha x\}$ fall into the same interval. This means that there are two integers $x_1 < x_2$ from the interval $\{0, 1, \dots, t\}$ such that

$$|\{\alpha x_2\} - \{\alpha x_1\}| < \frac{1}{t}.$$

Choose $q = x_2 - x_1$ (so that $1 \le q \le t$) and $p = \lfloor \alpha x_2 \rfloor - \lfloor \alpha x_1 \rfloor$ to get

$$|q\alpha - p| = |\alpha(x_2 - x_1) - (\lfloor \alpha x_2 \rfloor - \lfloor \alpha x_1 \rfloor)| = |\{\alpha x_2\} - \{\alpha x_1\}| < \frac{1}{t}.$$

This concludes the construction of the fraction p/q with required properties. □

As a corollary we deduce the following statement.

Theorem 6.4 (Dirichlet's theorem). *If α is a real* irrational *number then the inequality*

$$\left|\alpha - \frac{p}{q}\right| < \frac{1}{q^2}$$

has infinitely many solutions.

In fact, a stronger statement follows from the theory of continued fractions (see Section 4.4).

Proof. Taking $n = t$ in Theorem 6.3 we deduce that, for each $n = 1, 2, \ldots$, there is a fraction p_n/q_n such that $|\alpha - p_n/q_n| < 1/(nq_n)$ and $1 \le q_n \le n$. The two inequalities imply $|\alpha - p_n/q_n| \le 1/q_n^2$ since $nq_n \ge q_n^2$. Furthermore, $1/(nq_n) \to 0$ as $n \to \infty$, so that $|\alpha - p_n/q_n| \to 0$ as $n \to \infty$. Since the equality $|\alpha - p_n/q_n| = 0$ is not possible for irrational α, there are infinitely many elements in the sequence $\{p_n/q_n\}_{n=1}^{\infty}$ to accommodate the limiting relation. □

Observe that Dirichlet's theorem fails for rational $\alpha = a/b$, since for any fraction $p/q \ne \alpha$ we get

$$\left|\alpha - \frac{p}{q}\right| = \left|\frac{a}{b} - \frac{p}{q}\right| = \frac{|aq - bp|}{bq} \ge \frac{1}{bq} = \frac{1/b}{q} \ge \frac{1}{q^2}$$

whenever $q \ge b$. In fact, the estimate shows that in this case $|\alpha - p/q| \ge c/q$ for all positive integers q, where $c = 1/b$. Thus, we can state the following irrationality criterion.

Lemma 6.4. *If α is real and there are infinitely many fractions p/q such that $0 < |\alpha - p/q| = o(1/q)$ as $q \to \infty$, then α is irrational.*

Notice that the existence of infinitely many solutions p/q to the *diophantine* inequality $0 < |\alpha - p/q| = o(1/q)$ implies *a posteriori* a stronger conclusion (thanks to Dirichlet's theorem): there are infinitely many solutions p/q to the inequality $0 < |\alpha - p/q| < 1/q^2$.

Exercise 6.5. Show that for all integers p and $q \ge 2$,

$$\left|\sqrt{2} - \frac{p}{q}\right| > \frac{1}{4q^2}.$$

The next diophantine results are for the constant

$$e = \lim_{n\to\infty}\left(1 + \frac{1}{n}\right)^n = \sum_{k=0}^{\infty}\frac{1}{k!}$$
$$= 2.71828182845904523536028747135266249775724709369995\ldots.$$

It is classical that the partial sums of the latter series produce excellent rational approximations to the number, good enough for proving its irrationality (in view of Lemma 6.4, for example). We will establish somewhat stronger.

Theorem 6.5. (a) *The number e^2 is not a quadratic irrationality.*

(b) *The number e is not a quadratic irrationality.*

Proof. (a) Assume on the contrary that e^2 is a quadratic irrationality, so that there is a positive integer a and some integers b, c such that $ae^2 + be^{-2} + c = 0$.

As we know from Exercise 1.3, the exact power of prime 2 in $n!$ is equal to

$$\nu_2(n) = \left\lfloor \frac{n}{2} \right\rfloor + \left\lfloor \frac{n}{4} \right\rfloor + \left\lfloor \frac{n}{8} \right\rfloor + \cdots;$$

in particular, $\nu_2(2^m) = \nu_2(2^m + 1) = 2^{m-1} + \cdots + 2 + 1 = 2^m - 1$. It means that the rational number $2^n/n!$ written to the lowest terms as $2^{\alpha_n}/M_n$ has $\alpha_n = 1$ when $n = 2^m$ and $\alpha_n = 2$ if $n = 2^m + 1$ for some $m \in \mathbb{N}$. In what follows we choose and fix n of one of these forms to be sufficiently large.

From the Taylor series expansion with the remainder in the Lagrange form,

$$\begin{aligned} e^x &= \sum_{k=0}^{n} \frac{x^k}{k!} + \frac{x^{n+1}e^{\theta x}}{(n+1)!} \\ &= 1 + x + \frac{x^2}{2!} + \frac{x^3}{3!} + \cdots + \frac{x^n}{n!} \times \left(1 + \frac{xe^{\theta x}}{n+1}\right) \quad \text{for some } 0 < \theta < 1. \end{aligned}$$

Denote by β_+ and β_- the corresponding values of $e^{\theta x}/(n+1)$ for $x = 2$ and $x = -2$, respectively, so that

$$\begin{aligned} e^2 &= 1 + \cdots + \frac{2^{\alpha_k}}{M_k} + \cdots + \frac{2^{\alpha_n}}{M_n}(1 + 2\beta_+), \\ e^{-2} &= 1 - \cdots \pm \frac{2^{\alpha_k}}{M_k} \mp \cdots \pm \frac{2^{\alpha_n}}{M_n}(1 - 2\beta_-). \end{aligned}$$

Notice that $0 < \beta_+ < e^2/(n+1)$ and $0 < \beta_- < 1/(n+1)$; in particular, for all n sufficiently large we get

$$0 < a\beta_+ + |b|\beta_- < \frac{1}{8}.$$

Substituting the Taylor expansions into $ae^2 + be^{-2} + c = 0$ and multiplying the result by M_n (which is the 'odd' part of $n!$, so it is divisible by M_k for all $k < n$) we obtain

$$a \times 2^{\alpha_n} \times 2\beta_+ \mp b \times 2^{\alpha_n} \times 2\beta_- = d$$

for some integer d. By choosing $n = 2^m$ if $b \le 0$ and $n = 2^m + 1$ if $b > 0$ we get the left-hand side equal to $2^{\alpha_n+1}(a\beta_+ + |b|\beta_-)$, which is then a quantity in the range between 0 and 1 from the inequality above and $2^{\alpha_n+1} \in \{4,8\}$. Finally, the inequality $0 < d < 1$ is not possible for an integer d. The contradiction implies that e^2 is not a quadratic irrationality.

(b) If e were a quadratic irrationality then $ae+be^{-1} \in \mathbb{Z}$ for some $a,b \in \mathbb{Z}$, not simultaneously zero, so that $(ae+be^{-1})^2 = a^2e^2+b^2e^{-2}+2ab \in \mathbb{Z}$ meaning that e^2 is a quadratic irrationality. The latter is however excluded by part (a). □

6.3 Liouville's theorem on rational approximations of irrational algebraic numbers

The next result is what led Liouville (1844) to show for the first time that transcendental numbers exist.

Theorem 6.6 (Liouville's theorem). *Let α be algebraic number of degree $n \ge 2$. Then there exists a constant $c = c(\alpha)$ such that for any rational fraction p/q we have the inequality*

$$\left|\alpha - \frac{p}{q}\right| > \frac{c}{q^n}.$$

Proof. Denote

$$f(x) = a_nx^n + a_{n-1}x^{n-1} + \cdots + a_1x + a_0 \in \mathbb{Z}[x], \qquad a_n > 0,$$

the minimal polynomial of the algebraic number α with $\gcd(a_n, a_{n-1}, \dots, a_1, a_0) = 1$. Its factorisation over $\mathbb{C}$ reads

$$f(x) = a_n(x-\alpha) \times \prod_{j=2}^{n}(x-\alpha_j),$$

where $\alpha_2, \alpha_3, \dots, \alpha_n$ are algebraic conjugates of α. If $|\alpha - p/q| \ge 1$ then, clearly, $|\alpha - p/q| \ge 1/q^n$, so that the desired inequality holds with $c = 1$. In what follows we will therefore restrict ourselves to the case $|\alpha - p/q| < 1$.

From $|\alpha - p/q| < 1$ we deduce that $|p/q| < |\alpha| + 1$. This implies that

$$\begin{aligned}\left|f\left(\frac{p}{q}\right)\right| &= a_n\left|\alpha - \frac{p}{q}\right| \times \prod_{j=2}^{n}\left|\alpha_j - \frac{p}{q}\right| \le a_n\left|\alpha - \frac{p}{q}\right| \times \prod_{j=2}^{n}\left(|\alpha_j| + \left|\frac{p}{q}\right|\right) \\ &< a_n\left|\alpha - \frac{p}{q}\right| \times \prod_{j=2}^{n}(|\alpha_j| + |\alpha| + 1) \le a_n\left|\alpha - \frac{p}{q}\right| \times (2\lceil\overline{\alpha}\rceil + 1)^{n-1},\end{aligned}$$

where $\lceil\overline{\alpha}\rceil = \max\{|\alpha|, |\alpha_2|, \dots, |\alpha_n|\}$ is the so-called *house* of algebraic number α. On the other hand, $f(p/q) \neq 0$ since $f(x)$ is an irreducible polynomial of degree $n \geq 2$, hence

$$\left|f\left(\frac{p}{q}\right)\right| = \frac{|a_n p^n + a_{n-1}p^{n-1}q + \cdots + a_1 pq^{n-1} + a_0 q^n|}{q^n} \geq \frac{1}{q^n},$$

because all the numbers in the numerator are integral. Combining the two inequalities we obtain

$$\frac{1}{q^n} \leq \left|f\left(\frac{p}{q}\right)\right| < a_n\left|\alpha - \frac{p}{q}\right| \times \prod_{j=2}^{n}(2\lceil\overline{\alpha}\rceil + 1)$$

implying that

$$\left|\alpha - \frac{p}{q}\right| > \frac{c(\alpha)}{q^n}$$

with the choice

$$c(\alpha) = \frac{1}{a_n(2\lceil\overline{\alpha}\rceil + 1)^{n-1}} < 1.$$

□

Liouville's theorem implies that the diophantine inequality

$$0 < \left|\alpha - \frac{p}{q}\right| < \frac{c(\alpha)}{q^n}$$

does not have solutions in integers p and $q > 0$.

Lemma 6.5. *If for a real number α, for any $n \in \mathbb{N}$ the inequality*

$$0 < \left|\alpha - \frac{p}{q}\right| < \frac{1}{q^n}$$

has infinitely many solutions in integers p and $q > 0$, then α is transcendental.

Proof. Assume on the contrary that α is algebraic and choose $m \geq 2$ be its degree and $c = c(\alpha) > 0$ the corresponding constant from Theorem 6.6, so that $|\alpha - p/q| > c/q^m$ for all p/q. On the other hand, the hypothesis implies that there are infinitely many p/q satisfying $|\alpha - p/q| < 1/q^{m+1}$. In this infinite set we pick one with $q > 1/c$. Then

$$\left|\alpha - \frac{p}{q}\right| < \frac{1}{q^{m+1}} < \frac{c}{q^m},$$

which contradicts to the inequality above. □

Theorem 6.7. *The quantity*

$$\alpha = \sum_{k=1}^{\infty} \frac{1}{2^{k!}}$$

is transcendental.

Proof. Write the partial sums of the series

$$\frac{p_n}{q_n} = \sum_{k=1}^{n} \frac{1}{2^{k!}},$$

where $q_n = 2^{n!}$ and p_n are certain positive integers, $n = 1, 2, \dots$. Then

$$r_n = \alpha - \frac{p_n}{q_n} = \sum_{k=n+1}^{\infty} \frac{1}{2^{k!}}$$

satisfies $r_n > 0$ and

$$\begin{aligned} r_n &= \frac{1}{2^{(n+1)!}} \times \left(1 + \frac{1}{2^{(n+2)!-(n+1)!}} + \frac{1}{2^{(n+3)!-(n+1)!}} + \cdots\right) \\ &< \frac{1}{2^{(n+1)!}} \times \left(1 + \frac{1}{2} + \frac{1}{2^2} + \cdots\right) = \left(\frac{1}{2^{n!}}\right)^{n+1} \times 2 = \frac{2}{q_n^{n+1}} \\ &\le \frac{1}{q_n^n} \end{aligned}$$

for $n = 1, 2, \dots$. This means that, for each n, the inequality $0 < \alpha - p/q < 1/q^n$ has infinitely many solutions, namely,

$$\frac{p}{q} \in \left\{\frac{p_n}{q_n}, \frac{p_{n+1}}{q_{n+1}}, \frac{p_{n+2}}{q_{n+2}}, \dots\right\}.$$

It remains to apply the above corollary to Liouville's theorem — Lemma 6.5. □

6.4 Bounds for the value of polynomial at an algebraic point

In order to approach transcendence proofs more generally, we need some standard information about algebraic numbers; in particular, a suitable generalisation of Liouville's theorem. The next statement comes from algebra (so that we do not discuss its proof here).

Lemma 6.6. *Let G be a symmetric polynomial from the ring $\mathbb{Z}[x_1, \dots, x_n]$ of n variables and*

$$s_1 = x_1 + \cdots + x_n, \quad s_2 = x_1x_2 + \cdots + x_{n-1}x_n, \quad \dots, \quad s_n = x_1 \cdots x_n$$

elementary symmetric polynomials. Then there exists a polynomial $F \in \mathbb{Z}[y_1, \dots, y_n]$ of degree at most $\deg G$ *such that*

$$F(s_1, s_2, \dots, s_n) = G(x_1, \dots, x_n).$$

The *height* of a polynomial $P(x) = p_nx^n + \cdots + p_1x + p_0$ is the maximum of the absolute values of its coefficients, $H(P) = \max_{0\le k\le n} |p_k|$.

Exercise 6.6. The *height* $H(\alpha)$ of an algebraic number α is the height of its minimal polynomial from $\mathbb{Z}[x]$, whose coefficients do not possess a non-trivial common divisor (a so-called *primitive* polynomial). Establish bounds for the heights $H(\alpha+\beta)$ and $H(\alpha\beta)$ from above, where α and β are algebraic numbers, by means of $\deg\alpha$, $\deg\beta$, $H(\alpha)$ and $H(\beta)$.

Theorem 6.8. *Suppose that α is an algebraic number of degree $n \ge 1$. Then there exists a constant $c = c(\alpha) > 0$ such that for any polynomial $P(x)$ with integer coefficients, either $P(\alpha) = 0$ or*

$$|P(\alpha)| > \frac{c^k}{H^{n-1}},$$

where $k = \deg P$ is the degree of the polynomial and $H = H(P)$ its height.

Proof. Denote $f(x) = a_nx^n + \cdots + a_1x + a_0$ an (irreducible) minimal polynomial of α with integer coefficients and $a_n > 0$; its zeros are $\alpha_1 = \alpha$ and $\alpha_2, \dots, \alpha_n$. Take $P(x) \in \mathbb{Z}[x]$ an arbitrary polynomial. If $P(\alpha) = 0$ then there is nothing to prove, so we assume that $P(\alpha) \ne 0$. Then $P(\alpha_j) \ne 0$ for $j = 2, \dots, n$ by Lemma 6.2. Consider the symmetric polynomial

$$G(x_1, x_2, \dots, x_n) = P(x_1)P(x_2)\cdots P(x_n),$$

for which $G(x_1, \dots, x_n) = F(s_1, \dots, s_n)$ by Lemma 6.6. Then $G(\alpha_1, \dots, \alpha_n) = \prod_{j=1}^n P(\alpha_j) \ne 0$, while from Viète's theorem we obtain

$$G(\alpha_1, \dots, \alpha_n) = F\biggl(-\frac{a_{n-1}}{a_n}, \dots, (-1)^n\frac{a_0}{a_n}\biggr).$$

Since the coefficients of the polynomial F are integral and its degree is at most $\deg G = nk$, we conclude from the latter result that

$$G(\alpha_1, \dots, \alpha_n) = \frac{A}{a_n^{nk}}$$

for some nonzero integer A. Furthermore, for all $j = 2, \dots, n$ we have the estimates

$$|P(\alpha_j)| \le H(1 + |\alpha_j| + |\alpha_j|^2 + \cdots + |\alpha_j|^n) \le H(1 + |\alpha_j|)^k \le H(1 + \overline{|\alpha|})^k,$$

where $\overline{|\alpha|} = \max_{1\le j\le n} |\alpha_j|$ is the house of α, so that

$$|G(\alpha_1, \dots, \alpha_n)| \le |P(\alpha)|\, H^{n-1}(1 + \overline{|\alpha|})^{k(n-1)}.$$

Comparing this with

$$|G(\alpha_1,\dots,\alpha_n)| = \frac{|A|}{a_n^{nk}} \ge \frac{1}{a_n^{nk}}$$

we arrive at the desired estimate for $|P(\alpha)|$ with

$$c = c(\alpha) = \frac{1}{2a_n^n(1+\lceil\alpha\rceil)^{n-1}}. \qquad \square$$

Exercise 6.7. The *height* $H(P)$ of a polynomial in any number of variables is defined in exactly the same manner as in the single-variable case: it is the maximum of the absolute values of all coefficients of P.

Assume that α and β are algebraic numbers of degrees n and m, respectively. Prove the following extension of Theorem 6.8: There exists a constant $c = c(\alpha,\beta) > 0$ such that for any polynomial $P(x,y) \in \mathbb{Z}[x,y]$ in two variables of total degree $k \ge 1$ we either have $P(\alpha,\beta) = 0$ or

$$|P(\alpha,\beta)| > \frac{c^k}{H^{mn-1}}.$$

This exercise is clearly a part of some more general result, which we revisit again in Chapter 8 (without providing details of proof).

6.5 Transcendence of e

Our proof of transcendence of e relies on an analytical identity, due to Hermite, and a simple arithmetic fact.

Lemma 6.7 (Hermite's identity). *To a polynomial $f(x)$ of degree N with real coefficients, assign the polynomial*

$$F(x) = f(x) + f'(x) + \cdots + f^{(N)}(x),$$

where the sum is over all (nonzero) derivatives of $f(x)$. Then

$$F(0)e^x - F(x) = e^x \int_0^x f(t)e^{-t}\,\mathrm{d}t$$

for all $x > 0$.

Proof. The identity follows from the following repeated integration by parts:

$$\int_0^x f(t)e^{-t}\,\mathrm{d}t = \int_0^x f(t)\,\mathrm{d}(-e^{-t}) = f(0) - f(x)e^{-x} + \int_0^x f'(t)e^{-t}\,\mathrm{d}t = \cdots$$
$$= F(0) - F(x)e^{-x}. \qquad \square$$

Lemma 6.8. *If $g(x)$ is a polynomial with integer coefficients then so is the polynomial*

$$\frac{g^{(k)}(x)}{k!} = \frac{1}{k!}\frac{\mathrm{d}^k}{\mathrm{d}x^k}g(x)$$

for any $k \ge 0$.

Proof. Any such $g(x)$ is a $\mathbb{Z}$-linear combination of the monomials x^n, where $n = 0, 1, \dots$, hence it is sufficient to prove the statement for them. If $k > n$ then $(x^n)^{(k)}/k! \equiv 0$; otherwise we have

$$\frac{(x^n)^{(k)}}{k!} = \frac{n(n-1)\cdots(n-k+1)}{k!}x^{n-k} = \binom{n}{k}x^{n-k} \in \mathbb{Z}[x]. \qquad \square$$

Exercise 6.8. A polynomial $P(x) \in \mathbb{Q}[x]$ is said to be *integer-valued* if $P(k) \in \mathbb{Z}$ for all $k \in \mathbb{Z}$. Examples of such polynomials are $(x+2)(x-5)/2$, $x(x^2-1)/6$ and, more generally, the binomials

$$\binom{x}{n} = \frac{x(x-1)\cdots(x-n+1)}{n!} \quad \text{for } n = 1, 2, \dots, \qquad \binom{x}{0} = 1.$$

Prove that any integer-valued polynomial can be written as a linear combination of $\binom{x}{n}$ with integer coefficients.

Exercise 6.9. If $P(x)$ is an integer-valued polynomial of degree $m \ge 1$ and M is the least common multiple of $1, 2, \dots, m$, show that $MP'(x)$ is integer-valued.

Theorem 6.9 (Hermite). *The number e is transcendental.*

Proof. Assume, on the contrary, that e is algebraic and choose $\phi(x) = a_m x^m + \cdots + a_1 x + a_0$ to be its minimal polynomial with integer coefficients. Since $\phi(x)$ is irreducible, we have $a_0 \ne 0$. For a sufficiently large n, whose choice we will finalise later, consider the polynomial

$$f(x) = \frac{1}{(n-1)!}x^{n-1}(x-1)^n(x-2)^n\cdots(x-m)^n.$$

Apply Hermite's identity from Lemma 6.7 with this choice of $f(x)$ and for each $x = 0, 1, \dots, m$, and collect the results into a single linear combinations with the related coefficients $a_0, a_1, \dots, a_m$:

$$-\sum_{k=0}^{m} a_k F(k) = \sum_{k=0}^{m} a_k e^k \int_0^k f(t)e^{-t}\,\mathrm{d}t,$$

where we use the fact that $\sum_{k=0}^{m} a_k e^k F(0) = \phi(e)F(0) = 0$. Our aim is to show that the expression on the left-hand side of the formula is a nonzero integer when $n > \max\{m, |a_0|\}$ is a prime number, while the right-hand side tends to 0 as $n \to \infty$.

It follows from the definition of $f(x)$ that $f^{(j)}(0) = 0$ for $j = 0, 1, \dots, n-2$ and

$$f^{(n-1)}(0) = (x-1)^n(x-2)^n \cdots (x-m)^n\big|_{x=0} = (-1)^{mn}m!^n;$$

we also have $f^{(j)}(k) = 0$ for all $j = 0, 1, \dots, n-1$ and $k = 1, 2, \dots, m$. From Lemma 6.8 applied to the polynomial

$$g(x) = (n-1)!\, f(x) = x^{n-1}(x-1)^n(x-2)^n \cdots (x-m)^n$$

we conclude that the derivatives of $f(x)$ of orders $j = n, n+1, \dots$ are all divisible by $n = n!/(n-1)!$. In particular, $f^{(j)}(k) \in n\mathbb{Z}$ for all such j and $k = 0, 1, \dots, m$. Summarising our findings and using the definition of the polynomial $F(x)$ we see that

$$\sum_{k=0}^{m} a_k F(k) = (-1)^{mn}m!^n\, a_0 + nA$$

for some $A \in \mathbb{Z}$. If n is a prime number satisfying $n > m$ and $n > |a_0|$ then the integer $(-1)^{mn}m!^n\, a_0$ is not divisible by n. This means that whole expression is not divisible by n, so it is a nonzero integer. The implication is the estimate

$$\left|\sum_{k=0}^{m} a_k F(k)\right| \ge 1.$$

On the other hand, we have

$$|f(t)| < \frac{m^{n-1} \cdot m^n \cdots m^n}{(n-1)!} = \frac{m^{(m+1)n-1}}{(n-1)!}$$

for all t in the interval $0 \le t \le m$, so that

$$\begin{aligned}
\left|\sum_{k=0}^{m} a_k F(k)\right| &= \left|\sum_{k=0}^{m} a_k e^k \int_0^k f(t)e^{-t}\,\mathrm{d}t\right| \\
&< \frac{m^{(m+1)n-1}}{(n-1)!} \sum_{k=0}^{m} |a_k| e^k \int_0^k e^{-t}\,\mathrm{d}t \\
&< \frac{m^{(m+1)n-1}e^m}{(n-1)!} \sum_{k=0}^{m} |a_k| = \frac{c_0^n}{(n-1)!}c_1 < 1
\end{aligned}$$

for sufficiently large n, since $c_0^n/(n-1)! \to 0$ as $n \to \infty$. The two contradictory estimates we have obtained for $\sum_{k=0}^{m} a_k F(k)$ imply that e cannot be algebraic. □

6.6 Irrationality of π

A modification of the argument from the previous section allows one to prove the irrationality of π.

Lemma 6.9. *To a polynomial $f(x)$ with real coefficients, assign the polynomial*

$$F(x) = f(x) - f''(x) + f^{(4)}(x) - f^{(6)}(x) + \cdots.$$

Then

$$F(0) + F(\pi) = \int_0^\pi f(t) \sin t \, \mathrm{d}t.$$

Proof. The identity follows from integration of equality

$$\frac{\mathrm{d}}{\mathrm{d}x}\big(F'(x)\sin x - F(x)\cos x\big) = (F''(x) + F(x))\sin x = f(x)\sin x. \qquad \square$$

Theorem 6.10. *The number π is irrational.*

Proof. Assuming that π is rational, $\pi = a/b$ with $b > 0$, say, apply the identity of Lemma 6.9 to the polynomial

$$f(x) = \frac{b^n}{n!}x^n(\pi - x)^n = \frac{1}{n!}x^n(a - bx)^n,$$

where n is chosen sufficiently large. We have $f(0) = f'(0) = \cdots = f^{(n-1)}(0) = 0$, and also $f^{(j)}(0) \in \mathbb{Z}$ for $j \geq n$ from Lemma 6.8 applied to the polynomial $g(x) = x^n(a - bx)^n \in \mathbb{Z}[x]$. This means that $f^{(j)}(0) \in \mathbb{Z}$ for all $j \geq 0$. Because of the symmetry $f(\pi - x) = f(x)$, we get $f^{(j)}(\pi - x) = (-1)^j f^{(j)}(x)$, hence $f^{(j)}(\pi) = (-1)^j f^{(j)}(0) \in \mathbb{Z}$ for all $j \geq 0$. Combining this we conclude that both $F(0)$ and $F(\pi)$ are integers, hence

$$\int_0^\pi f(t) \sin t \, \mathrm{d}t = F(0) + F(\pi) \in \mathbb{Z}.$$

On the other hand, the integrand $f(t)\sin t$ is clearly positive on the interval $0 < t < \pi$ and possesses the estimate

$$f(t)\sin t \leq f(t) < \frac{b^n \pi^{2n}}{n!}$$

there; therefore,

$$0 < \int_0^\pi f(t)\sin t \, \mathrm{d}t < \frac{b^n \pi^{2n+1}}{n!} < 1$$

for all large n, since $b^n\pi^{2n+1}/n! \to 0$ as $n \to \infty$. As no integer exists between 0 and 1, our estimates lead to contradiction. Thus, π cannot be rational. $\square$

6.7 Newton's interpolating series for $\sin \pi z$

Assume that a function $f(z)$ is analytic in a domain $D \subset \mathbb{C}$, while $z_1, \dots, z_n$ is a fixed collection of points in D, with possible repetitions. Define $F_0(\zeta) = 1$ and

$$F_k(\zeta) = (\zeta - z_k)F_{k-1}(\zeta) = (\zeta - z_1)\cdots(\zeta - z_k) \quad \text{for } k = 1, \dots, n.$$

Multiplying the both sides of elementary identity

$$\frac{1}{\zeta - z}\left(1 - \frac{z - z_k}{\zeta - z_k}\right) = \frac{1}{\zeta - z_k}, \quad \text{where } k = 1, \dots, n,$$

by $F_{k-1}(z)/F_{k-1}(\zeta)$ we arrive at

$$\frac{1}{\zeta - z}\left(\frac{F_{k-1}(z)}{F_{k-1}(\zeta)} - \frac{F_k(z)}{F_k(\zeta)}\right) = \frac{F_{k-1}(z)}{F_k(\zeta)}, \quad \text{where } k = 1, \dots, n.$$

Summing them over k we get

$$\frac{1}{\zeta - z} - \frac{F_n(z)}{F_n(\zeta)(\zeta - z)} = \sum_{k=1}^{n} \frac{F_{k-1}(z)}{F_k(\zeta)},$$

equivalently,

$$\frac{1}{\zeta - z} = \sum_{k=1}^{n} \frac{F_{k-1}(z)}{F_k(\zeta)} + \frac{F_n(z)}{F_n(\zeta)(\zeta - z)}.$$

Now take a simple closed contour C within the domain D, which encloses all the points $z_1, \dots, z_n$ and a point z. Multiplying the identity obtained by $f(\zeta)/(2\pi i)$ and integrating the result along C we obtain

$$\begin{aligned} f(z) &= \frac{1}{2\pi i}\oint_C \frac{f(\zeta)}{\zeta - z}\,\mathrm{d}\zeta \\ &= \sum_{k=1}^{n} F_{k-1}(z)\,\frac{1}{2\pi i}\oint_C \frac{f(\zeta)}{F_k(\zeta)}\,\mathrm{d}\zeta + \frac{1}{2\pi i}\oint_C \frac{F_n(z)f(\zeta)}{F_n(\zeta)(\zeta - z)}\,\mathrm{d}\zeta. \end{aligned}$$

The resulting formula is known as Newton's interpolation formula with interpolation nodes $z_1, \dots, z_n$.

Lemma 6.10 (Newton's interpolation). *In the above notation, denote*

$$A_{k-1} = \frac{1}{2\pi i}\oint_C \frac{f(\zeta)}{F_k(\zeta)}\,\mathrm{d}\zeta \quad \textit{for } k = 1, \dots, n$$

and

$$R_n(z) = \frac{1}{2\pi i}\oint_C \frac{F_n(z)f(\zeta)}{F_n(\zeta)(\zeta - z)}\,\mathrm{d}\zeta.$$

Then for z within the contour C,

$$f(z) = \sum_{k=0}^{n-1} A_k F_k(z) + R_n(z).$$

From now on, think of a collection of points $z_1, z_2, \dots$ in D being infinite. Assuming that

$$R_n(z) = \frac{1}{2\pi i} \oint_C \frac{F_n(z)f(\zeta)}{F_n(\zeta)(\zeta - z)}\,d\zeta \to 0 \quad \text{as } n \to \infty$$

for all z from a domain $D_0 \subset D$, we then obtain the so-called *Newton's interpolating series*

$$f(z) = \sum_{k=0}^{\infty} A_k F_k(z) = \sum_{k=0}^{\infty} A_k (z - z_1) \cdots (z - z_k).$$

If $f(z)$ does not happen to be a polynomial, the latter equality involves infinitely many terms, hence $A_k \neq 0$ for infinitely many indices k.

We will next write Newton's interpolating series for the function $f(z) = \sin \pi z$ using the collection of nodes $z_1, z_2, \dots$ as follows. We fix a positive integer m and define $z_k = k$ for $k = 1, \dots, m$ and then $z_{m+k} = z_k$ for all $k \geq 1$. Choose and fix an arbitrary real $R > m$, so that all nodes are within the disc of radius R, and consider the remainder

$$R_n(z) = \frac{1}{2\pi i} \oint_C \frac{(z - z_1) \cdots (z - z_n) \sin \pi\zeta}{(\zeta - z_1) \cdots (\zeta - z_n)(\zeta - z)}\,d\zeta$$

for $|z| \leq R$ and $n > 2R$, choosing the contour C to be the circle $|\zeta| = n$. First, $|z - z_k| \leq |z| + |z_k| \leq R + m$ for $k = 1, \dots, n$. Second, on the circle $|\zeta| = n > 2R > 2m$ we have

$$|\zeta - z_k| \geq |\zeta| - |z_k| \geq n - m > \frac{n}{2} \quad \text{for } k = 1, \dots, n,$$

and $|\zeta - z| \geq |\zeta| - |z| \geq n - R > n/2$. Third, on the same circle we get

$$|\sin \pi\zeta| = \left| \frac{e^{\pi i\zeta} - e^{-\pi i\zeta}}{2i} \right| \leq e^{\pi|\zeta|} = e^{\pi n}.$$

Combining all these estimates, we deduce that

$$\begin{aligned} |R_n(z)| &\leq \frac{1}{2\pi} \oint_{|\zeta|=n} \frac{(R+m)^n e^{\pi n}}{(n/2)^{n+1}}\,d\zeta \\ &= \frac{2^{n+1}(R+m)^n e^{\pi n}}{n^n} \to 0 \quad \text{as } n \to \infty, \end{aligned}$$

since R and m are fixed. The quantity R can be chosen from the beginning arbitrary large, to include a given z inside the disc of radius R, so that the interpolating series is valid for $\sin \pi z$ with any choice of complex z.

Lemma 6.11 (Newton's interpolation for $\sin\pi z$). *Define the collection $z_1, z_2, \ldots$ by $z_k = k$ for $k = 1, \ldots, m$ and, recursively, $z_{m+k} = z_k$ for all $k \ge 1$. Take $z \in \mathbb{C}$. Then*

$$\sin\pi z = \sum_{n=0}^{\infty} A_n(z - z_1)\cdots(z - z_n),$$

where for $n > 2\max\{m, |z|\}$ the coefficients A_n satisfy

$$|A_n| < \exp(-n\ln n + 5n).$$

Proof. It remains to show the estimates for

$$A_n = \frac{1}{2\pi i}\oint_{|\zeta|=n} \frac{\sin\pi\zeta}{(\zeta - z_1)\cdots(\zeta - z_{n+1})}\,\mathrm{d}\zeta.$$

With the help of the bounds above we deduce that

$$|A_n| \le \frac{1}{2\pi}\frac{e^{\pi n}}{(n/2)^{n+1}}\,2\pi n = \frac{e^{\pi n+(n+1)\ln 2}}{n^n} < \frac{e^{5n}}{n^n},$$

the required bound. $\square$

6.8 Transcendence of π

For the periodic collection $z_1, z_2, \ldots$ defined by $z_k = k$ for $k = 1, \ldots, m$ and $z_{m+k} = z_k$ for all $k \ge 1$, we can write

$$(z - z_1)\cdots(z - z_{n+1}) = \prod_{k=1}^{m}(z - k)^{r_k+1},$$

where the integers r_k for $k \ge 1$ satisfy the hypotheses

$$r_1 + \cdots + r_m + m = n + 1,$$
$$r_1 - 1 \le r_m \le r_{m-1} \le \cdots \le r_2 \le r_1 \le \frac{n}{m}.$$

Then the interpolation coefficients A_n in Lemma 6.11 can be given by

$$A_n = \frac{1}{2\pi i}\oint_{|\zeta|=N} \frac{\sin\pi\zeta}{(\zeta - 1)^{r_1+1}\cdots(\zeta - m)^{r_m+1}}\,\mathrm{d}\zeta$$

for (any) choice of $N > 2m$. Denote $r = r_1 = \max_{1\le k\le m}\{r_k\}$ and M is the least common multiple of the numbers $1, 2, \ldots, m$.

Lemma 6.12. *For each $n \ge 0$, there exists a polynomial $P_n(x) \in \mathbb{Z}[x]$ of degree at most r and of height not exceeding $r!\,(2M)^n$ such that*

$$M^{n-1}r!\,A_n = P_n(\pi).$$

Proof. Use Cauchy's residue sum theorem to write

$$A_n = \sum_{k=1}^{m} \frac{1}{2\pi i} \oint_{\Gamma_k} \frac{\sin \pi\zeta}{(\zeta-1)^{r_1+1}\cdots(\zeta-m)^{r_m+1}}\,\mathrm{d}\zeta,$$

where Γ_k are circles $|\zeta - k| = \frac{1}{2}$ bypassed in the positive direction. Develop the function $\sin \pi\zeta$ into its Taylor series at $\zeta = k$:

$$\begin{aligned}\sin \pi\zeta &= \sin(\pi k + \pi(\zeta - k)) = (-1)^k \sin \pi(\zeta - k)\\ &= \sum_{j=0}^{\infty} \frac{(-1)^{j+k}\pi^{2j+1}}{(2j+1)!}(\zeta-k)^{2j+1}\\ &= \sum_{0\le j\le (r_k-1)/2} \frac{(-1)^{j+k}\pi^{2j+1}}{(2j+1)!}(\zeta-k)^{2j+1} + R_k(\zeta),\end{aligned}$$

where $R_k(\zeta)$ is an entire function with the zero of order at least r_k+1 at $\zeta = k$, where $k = 1, \ldots, m$. Therefore,

$$\frac{1}{2\pi i}\oint_{\Gamma_k} \frac{R_k(\zeta)}{(\zeta-1)^{r_1+1}\cdots(\zeta-m)^{r_m+1}}\,\mathrm{d}\zeta = 0,$$

hence

$$\begin{aligned}&\frac{1}{2\pi i}\oint_{\Gamma_k} \frac{\sin \pi\zeta}{(\zeta-1)^{r_1+1}\cdots(\zeta-m)^{r_m+1}}\,\mathrm{d}\zeta\\ &\quad= \sum_{0\le j\le (r_k-1)/2} \frac{(-1)^{j+k}\pi^{2j+1}}{(2j+1)!}\,\frac{1}{2\pi i}\oint_{\Gamma_k} \frac{(\zeta-k)^{2j+1}}{(\zeta-1)^{r_1+1}\cdots(\zeta-m)^{r_m+1}}\,\mathrm{d}\zeta.\end{aligned}$$

Denote, for each $k = 1, \ldots, m$,

$$a_{kj} = \frac{1}{2\pi i}\oint_{\Gamma_k} \frac{(\zeta-k)^{2j+1}}{(\zeta-1)^{r_1+1}\cdots(\zeta-m)^{r_m+1}}\,\mathrm{d}\zeta,$$

where $0 \le j \le (r_k - 1)/2$.

We shall now check that the numbers a_{kj} are rational such that $M^{n-1}a_{kj} \in \mathbb{Z}$. Each a_{kj} is a residue of the integrand at $\zeta = k$, that is, the coefficient of $(\zeta - k)^{-1}$ in the Laurent expansion of the rational function

$$\frac{(\zeta-k)^{2j+1}}{(\zeta-1)^{r_1+1}\cdots(\zeta-m)^{r_m+1}}$$

at $\zeta = k$. For integers $s \ne k$ in the range $1 \le s \le m$, we have

$$\frac{1}{\zeta - s} = \frac{1}{k-s} \times \frac{1}{1 - \dfrac{\zeta-k}{s-k}} = \frac{1}{k-s}\sum_{\ell=0}^{\infty}\left(\frac{\zeta-k}{s-k}\right)^{\ell} = -\sum_{\ell=0}^{\infty}\frac{(\zeta-k)^{\ell}}{(s-k)^{\ell+1}}$$

implying after r_s-repeated differentiation that

$$\frac{1}{(\zeta-s)^{r_s+1}} = (-1)^{r_s+1}\sum_{\ell=r}^{\infty}\binom{\ell}{r_s}\frac{(\zeta-k)^{\ell-r_s}}{(s-k)^{\ell+1}}$$
$$= (-1)^{r_s+1}\sum_{\ell=0}^{\infty}\binom{\ell+r_s}{\ell}\frac{(\zeta-k)^{\ell}}{(s-k)^{\ell+r_s+1}}.$$

(If $r_s = -1$ then the Laurent series expansion at $\zeta = k$ is simply $1/((\zeta - s)^{r_s+1} = 1$.) Therefore, the coefficient of $(\zeta-k)^{-1}$ in the Laurent expansion of

$$\frac{(\zeta-k)^{2j+1}}{(\zeta-1)^{r_1+1}\cdots(\zeta-m)^{r_m+1}} = \frac{1}{(\zeta-1)^{r_1+1}\cdots(\zeta-k)^{r_k-2j}\cdots(\zeta-m)^{r_m+1}}$$

at $\zeta = k$ is equal to a linear combination with integral coefficients of products

$$\prod_{\substack{s=1\\ s\neq k}}^{m}\frac{1}{(s-k)^{\ell_s+r_s+1}},$$

for which $\ell_1+\cdots+\ell_{k-1}+\ell_{k+1}+\cdots+\ell_m = r_k-(2j+1)$. Since $|s-k|$ is an integer between 1 and m while M denotes the least common multiple of all integers in the range, we have $M/(s-k)\in\mathbb{Z}$ and

$$\prod_{\substack{s=1\\ s\neq k}}^{m}\left(\frac{M}{s-k}\right)^{\ell_s+r_s+1}\in\mathbb{Z}.$$

Here

$$\sum_{\substack{s=1\\ s\neq k}}^{m}(\ell_s+r_s+1) = (m-1)+\sum_{s=1}^{m}r_s-(2j+1) = n-(2j+1)\le n-1,$$

so that indeed $M^{n-1}a_{kj}\in\mathbb{Z}$.

We now summarise our findings as follows: the quantity

$$r!\,M^{n-1}A_n = \sum_{k=1}^{m}\sum_{0\le j\le(r_k-1)/2}(-1)^{j+k}M^{n-1}a_{kj}\frac{r!}{(2j+1)!}\pi^{2j+1}$$

is a polynomial $P_n(x)\in\mathbb{Z}[x]$ evaluated at $x=\pi$, as required. It only remains to estimate the height of the polynomial from above. For each a_{kj} we use the defining integral representation and the fact that $|\zeta-k| = \frac{1}{2}$ and $|\zeta-s|\ge\frac{1}{2}$ for $s\neq k$ on the contour Γ_k:

$$|a_{kj}|\le\frac{1}{2\pi}\times\pi\times\frac{1}{(1/2)^{n-j}}\le 2^{n-1}$$

for $0\le j\le(r_k-1)/2$ and $1\le k\le m$. Thus, the absolute values of integer coefficients of $P_n(x)$ do not exceed

$$m\times r!\,2^{n-1}M^{n-1}\le r!\,(2M)^n. \qquad \square$$

Theorem 6.11 (Lindemann). *The number π is transcendental.*

Proof. Assume on the contrary that π is algebraic, of degree $m-1$. With this m consider the expansion of $\sin \pi z$ in the Newton interpolation series,

$$\sin \pi z = \sum_{n=0}^{\infty} A_n (z - z_1) \cdots (z - z_n),$$

where $|A_n| < \exp(-n \ln n + 5n)$ for $n > 2 \max\{m, |z|\}$. We choose n sufficiently large and consider the polynomial $P_n(x)$ constructed in Lemma 6.12, whose degree $\deg P_n \le r$ and height $H = H(P_n) \le r!\,(2M)^n$. The letters c, c_1, c_2 below are used to denote constants that only depend on m (recall that $m-1$ is the degree of algebraic number π). It follows from Theorem 6.8 that either $P_n(\pi) = 0$ or

$$\begin{aligned} |P_n(\pi)| &\ge \frac{c^r}{H^{m-2}} = e^{r \ln c - (m-2) \ln H} \\ &\ge e^{-r|\ln c| - (m-2)(r \ln r + n \ln(2M))} > \exp\left(-\frac{m-2}{m} n \ln n - c_1 n \right), \end{aligned}$$

where we used $r \le n/m$. On the other hand, from the estimate for A_n and the fact that $P_n(\pi) = r!\, M^{n-1} A_n$ we find out that

$$\begin{aligned} |P_n(\pi)| &< \exp(-n \ln n + 5n + (n-1) \ln M + r \ln r) \\ &< \exp\left(-\frac{m-1}{m} n \ln n + c_2 n \right). \end{aligned}$$

Comparing the two estimates for $|P_n(\pi)|$ we conclude that they are incompatible for all n sufficiently large, $n \ge N$. This means that we have $P_n(\pi) = 0$ for all $n \ge N$, so that $A_n = 0$ for all such n, hence $\sin \pi z$ is a polynomial. At the same time, it is not (for example, because it has infinitely many zeros on the real line). The contradiction we arrived at proves that π is transcendental. □

Exercise 6.10. Prove that the function $f(z) = \sin \pi z$ is transcendental. In other words, show that there is no polynomial

$$P(z, x) = \sum_{j=0}^{n} P_j(z) x^j$$

with rational-function coefficients $P_j(z)$ such that $P(z, f(z)) = 0$ identically in z.

Chapter notes

The proof of Theorem 6.5 is based on the argument of Liouville (1840).

Dirichlet's theorem (Theorem 6.4) and Liouville's theorem (Theorem 6.6) suggest investigating in general the question about how well a real number (not necessarily algebraic!) can be approximated by rationals. A standard way for measuring the quality of rational approximations to $\alpha \in \mathbb{R}$ is in terms of the *irrationality exponent* $\mu(\alpha)$, which is defined as the supremum of $\mu > 0$ for which the inequality

$$0 < \left|\alpha - \frac{p}{q}\right| < \frac{1}{q^{\mu}}$$

has infinitely many solutions in integers p and $q \neq 0$. Then Dirichlet's theorem translates into the inequality $\mu(\alpha) \geq 2$ for all *irrational* real α, while Liouville's theorem tells that $\mu(\alpha) \leq n$ for *algebraic* real α of degree at most n. The last result is in fact best possible for quadratic irrationalities (when $n = 2$) — this follows from Theorems 4.4 and 4.9, but not for $n > 2$. Roth's celebrated theorem [68] proves that $\mu(\alpha) = 2$ for all algebraic $\alpha \in \mathbb{R}$, and with a simple argument from the measure theory one can show that the irrationality exponent 2 is for almost every real number (in the sense of Lebesgue measure). But it is not always 2! Already the Liouville number α in Theorem 6.7 has $\mu(\alpha) = \infty$, as follows from the inequalities established in its proof. While showing that $\mu(e) = 2$ is considerably simple (see, for example, [14, Section 2.12]), estimating the irrationality exponent from above of other 'interesting' *irrational* constants is a competitive business. The latest record bound [85] set for the number π is $\mu(\pi) \leq 7.103205\ldots$ (though we expect it to be 2).

Chapter 7

Irrationality of zeta values

In this chapter we discuss arithmetic properties of the values of Riemann's zeta function $\zeta(s)$ at integers $s = 2, 3, 4, \dots$.

As we already know from Section 3.3 (see Proposition 3.7), the values of Riemann's zeta function $\zeta(s)$ at positive even integers $s = 2k$ happen to be rational multiples of π^{2k}, where $k = 1, 2, \dots$. Now, using the fact that π is a transcendental number (Theorem 6.11) we end up with the following immediate corollary.

Theorem 7.1. *The value $\zeta(2k)$ of Riemann's zeta function at an even integer $s = 2k$ is an irrational and transcendental number.*

Much less is known on the arithmetic nature of the zeta values at odd integers $s = 3, 5, 7, \dots$: in 1978, Apéry proved [3,64] the irrationality of the number

$$\zeta(3) = \sum_{n=1}^{\infty} \frac{1}{n^3}$$
$$= 1.2020569031595942853997381615114499907649862923404 9\dots,$$

and there are more recent but partial linear independence results of Rivoal [67] and others. Rivoal's theorem [67] settles the infinitude of the set of irrational numbers among $\zeta(3), \zeta(5), \zeta(7), \dots$. Conjecturally, each of these numbers is transcendental, and a complete answer to the above-stated question, about polynomial relations over $\mathbb{Q}$ for the values of $\zeta(s)$ with $s \geq 2$ integer, looks rather simple.

Conjecture 7.1. The numbers

$$\pi,\ \zeta(3),\ \zeta(5),\ \zeta(7),\ \zeta(9),\ \dots$$

are algebraically independent over $\mathbb{Q}$.

This conjecture may be regarded as a mathematical folklore. It seems to be unattainable by the present methods. Below we give a proof of Apéry's result and then discuss a partial result about the irrationality of other odd zeta values.

7.1 Arithmetic of linear forms in 1 and $\zeta(3)$

For $n = 0, 1, 2, \dots$, consider the rational function

$$Q_n(t) = \frac{(t-1)(t-2)\cdots(t-n)}{t(t+1)(t+2)\dots(t+n)} = \frac{\prod_{j=1}^{n}(t-j)}{\prod_{j=0}^{n}(t+j)}.$$

As the degree of its numerator is less than that of its denominator, its partial-fraction decomposition assumes the form

$$Q_n(t) = \sum_{k=0}^{n} \frac{c_k}{t+k}.$$

Lemma 7.1. *The coefficients c_k, where $k = 0, 1, \dots, n$, are integers.*

Proof. The standard procedure of expanding a rational fraction into the sum of partial fractions leads to

$$\begin{aligned} c_k = Q_n(t)(t+k)\big|_{t=-k} &= \frac{\prod_{j=1}^{n}(t-j)}{\prod_{j=0}^{k-1}(t+j) \times \prod_{j=k+1}^{n}(t+j)}\Bigg|_{t=-k} \\ &= \frac{(-1)^n \prod_{j=1}^{n}(k+j)}{(-1)^k k! \times (n-k)!} = (-1)^{n-k}\binom{n+k}{n}\binom{n}{k} \in \mathbb{Z} \end{aligned}$$

for $k = 0, 1, \dots, n$. □

Exercise 7.1. Show that the coefficients in the partial-fraction decompositions of the rational functions

$$\frac{n!}{\prod_{j=0}^{n}(t+j)}, \quad \frac{\prod_{j=1}^{n}(t+n+j)}{\prod_{j=0}^{n}(t+j)}, \quad \frac{2^{2n}\prod_{j=1}^{n}(t+\frac{1}{2}-j)}{\prod_{j=0}^{n}(t+j)},$$
$$\frac{2^{2n}\prod_{j=1}^{n}(t-\frac{1}{2}+j)}{\prod_{j=0}^{n}(t+j)} \quad \text{and} \quad \frac{2^{2n}\prod_{j=1}^{n}(t+n-\frac{1}{2}+j)}{\prod_{j=0}^{n}(t+j)}$$

possess the same property as displayed in Lemma 7.1.

Denote $d_n = \operatorname{lcm}(1, 2, \dots, n)$. The asymptotics of this quantity is controlled by the prime number theorem.

Lemma 7.2. *We have*

$$\lim_{n\to\infty} \frac{\ln d_n}{n} = 1;$$

in other words, d_n grows with n like $e^{n+o(n)}$ as $n \to \infty$.

Proof. It is not hard to see that d_n is a product over primes $p \le n$ entering with exponent k such that $p^k \le n$. This means that

$$\ln d_n = \sum_{p^k \le n} k \ln p = \sum_{p \le n} \left\lfloor \frac{\ln n}{\ln p} \right\rfloor \ln p = \psi(n),$$

where ψ is Chebyshev's function from Section 2.6. Thus, the required asymptotics follows from Lemma 2.10 and Theorem 2.8. □

We now turn our attention to the rational function

$$R_n(t) = Q_n(t)^2 = \frac{\prod_{j=1}^{n}(t-j)^2}{\prod_{j=0}^{n}(t+j)^2},$$

which plays a special role in our construction of rational approximations to $\zeta(3)$.

Lemma 7.3. *The rational coefficients in the partial-fraction decomposition*

$$R_n(t) = \sum_{k=0}^{n} \left(\frac{a_k}{(t+k)^2} + \frac{b_k}{t+k} \right)$$

satisfy the inclusions $a_k \in \mathbb{Z}$ and $d_n b_k \in \mathbb{Z}$ for $k = 0, 1, \ldots, n$.

Proof. Notice that a decomposition of rational function into the sum of partial fractions is unique. Use the partial-fraction expansion of Lemma 7.1,

$$\begin{aligned} R_n(t) = \left(\sum_{k=0}^{n} \frac{c_k}{t+k} \right)^2 &= \sum_{k=0}^{n} \left(\frac{c_k}{t+k} \right)^2 + \sum_{\substack{k=0 \\ k\ne l}}^{n} \sum_{l=0}^{n} \frac{c_k c_l}{(t+k)(t+l)} \\ &= \sum_{k=0}^{n} \frac{c_k^2}{(t+k)^2} + \sum_{\substack{k=0 \\ k\ne l}}^{n} \sum_{l=0}^{n} \frac{c_k c_l}{l-k} \left(\frac{1}{t+k} - \frac{1}{t+l} \right), \end{aligned}$$

implying

$$a_k = c_k^2 = \binom{n+k}{n}^2 \binom{n}{k}^2 \quad \text{and} \quad b_k = 2c_k \sum_{\substack{l=0 \\ l\ne k}}^{n} \frac{c_l}{l-k} \quad \text{for } k = 0, 1, \ldots, n.$$

Since $|l-k| \le n$ in the latter sum, the resulting formulae for a_k and b_k give us grounds for the required inclusions. □

Finally, consider the sequence

$$r_n = -\sum_{m=1}^{\infty} \frac{\mathrm{d}R_n}{\mathrm{d}t}\bigg|_{t=m}. \tag{7.1}$$

Lemma 7.4. *For each $n = 0, 1, 2, \dots$, the quantity r_n can be represented in the form $r_n = q_n\zeta(3) - p_n$ with $q_n \in \mathbb{Z}$ and $d_n^3 p_n \in \mathbb{Z}$.*

Proof. We have

$$\begin{aligned} r_n &= \sum_{m=1}^{\infty}\sum_{k=0}^{n}\left(\frac{2a_k}{(t+k)^3} + \frac{b_k}{(t+k)^2}\right)\bigg|_{t=m} \\ &= 2\sum_{k=0}^{n} a_k \sum_{m=1}^{\infty} \frac{1}{(m+k)^3} + \sum_{k=0}^{n} b_k \sum_{m=1}^{\infty} \frac{1}{(m+k)^2} \\ &= 2\sum_{k=0}^{n} a_k\left(\zeta(3) - \sum_{\ell=1}^{k}\frac{1}{\ell^3}\right) + \sum_{k=0}^{n} b_k\left(\zeta(2) - \sum_{\ell=1}^{k}\frac{1}{\ell^2}\right). \end{aligned}$$

Observe that

$$\sum_{k=0}^{n} b_k = \sum_{k=0}^{n} \operatorname{Res}_{t=-k} R_n(t) = -\operatorname{Res}_{t=\infty} R_n(t)$$

by the residue sum theorem, and $\operatorname{Res}_{t=\infty} R_n(t) = 0$ because $R_n(t) = O(t^{-2})$ as $t \to \infty$. It follows that $r_n = q_n\zeta(3) - p_n$, where

$$q_n = 2\sum_{k=0}^{n} a_k \in \mathbb{Z} \quad \text{and} \quad p_n = 2\sum_{k=0}^{n} a_k \sum_{\ell=1}^{k}\frac{1}{\ell^3} + \sum_{k=0}^{n} b_k \sum_{\ell=1}^{k}\frac{1}{\ell^2}.$$

Finally, the inclusions $d_n^3 p_n \in \mathbb{Z}$ follow from the explicit formula for p_n and Lemma 7.4. □

The numbers

$$\frac{1}{2}q_n = \sum_{k=0}^{n}\binom{n+k}{n}^2\binom{n}{k}^2$$

showing up as the coefficients of $\zeta(3)$ in the linear form are known as the Apéry numbers.

Exercise 7.2 (Apéry's recursion). (a) Verify that

$$r_0 = 2\zeta(3) \quad \text{and} \quad r_1 = 10\zeta(3) - 12.$$

(b) Define $S_n(t) = s_n(t)R_n(t)$, where $s_n(t) = 4(2n+1)(-2t^2+t+(2n+1)^2)$. Check that

$$(n+1)^3R_{n+1}(t) - (2n+1)(17n^2+17n+5)R_n(t) + n^3R_{n-1}(t) = S_n(t+1) - S_n(t)$$

for $n = 1, 2, \ldots$.

(c) Using part (b), show that the sequences $\{r_n\}_{n=0}^{\infty}$, $\{q_n\}_{n=0}^{\infty}$ and $\{p_n\}_{n=0}^{\infty}$ satisfy the same(!) recurrence relation

$$(n+1)^3r_{n+1}-(2n+1)(17n^2+17n+5)r_n+n^3r_{n-1} = 0 \quad \text{for } n = 1, 2, \ldots. \tag{7.2}$$

The argument for deducing Apéry's recursion (7.1) from the exercise is known as creative telescoping [81, 84].

7.2 Apéry's theorem

It remains to determine the growth of the linear forms $r_n = q_n\zeta(3) - p_n$ constructed in Lemma 7.4 as $n \to \infty$. For that we will use Stirling's formula for the gamma function and apply the saddle-point method from analysis.

Lemma 7.5. *For the sum r_n in (7.1) the following integral representation is valid:*

$$r_n = \frac{1}{2\pi i}\int_{C-i\infty}^{C+i\infty}\left(\frac{\pi}{\sin \pi t}\right)^2 R_n(t)\,\mathrm{d}t,$$

where in the contour of integration $\operatorname{Re} t = C$ *one can take any C in the interval* $0 < C < n+1$.

Proof. Fix $N > n$ and consider the rectangular contour (positively oriented) with vertices at $C \pm iN$ and $N + \frac{1}{2} \pm iN$. The function $\pi/\sin \pi t$ is bounded on the sides $\operatorname{Im} t = \pm N$ of the contour: for example, for $t = x - iN$, we find that

$$\frac{\pi}{|\sin \pi t|} = \frac{2\pi}{|e^{\pi(N+ix)} - e^{-\pi(N+ix)}|} = \frac{2\pi e^{-\pi N}}{|e^{\pi ix} - e^{-\pi(2N+ix)}|} \le \frac{2\pi e^{-\pi N}}{1 - e^{-2\pi N}} < 4\pi e^{-\pi N},$$

and the same bound is valid for $t = x + iN$. It is also bounded on the side $\operatorname{Re} t = N + \frac{1}{2}$ of the rectangle: when $t = N + \frac{1}{2} + iy$, we get

$$\frac{\pi}{|\sin \pi s|} = \frac{\pi}{\cosh \pi y} < 2\pi e^{-\pi|y|} \le 2\pi.$$

The function $R_n(t)$ is $O(t^{-2})$ as $t \to \infty$, hence it is $O(N^{-2})$ on the three sides of the contour. By performing the limit as $N \to \infty$, it follows that the complex integral

$$-\frac{1}{2\pi i}\int_{C-i\infty}^{C+i\infty}\left(\frac{\pi}{\sin \pi t}\right)^2 R_n(t)\,\mathrm{d}t$$

equals the sum of the residues of the integrand at the poles $t = m$, where m runs over the integers satisfying $m > C$. Since

$$\left(\frac{\pi}{\sin \pi t}\right)^2 = \frac{1}{(t-m)^2} + O(1)$$

and

$$R_n(t) = R_n(m) + R_n'(m)(t-m) + O\big((t-m)^2\big)$$

as $t \to m$, we conclude that

$$\operatorname{Res}_{t=m}\left(\frac{\pi}{\sin \pi t}\right)^2 R_n(t) = R_n'(m).$$

It remains to notice that $R_n'(m) = 0$ for $m = 1, 2, \dots, n$, so that

$$\sum_{m>C}\operatorname{Res}_{t=m}\left(\frac{\pi}{\sin \pi t}\right)^2 R_n(t) = \sum_{m>C} R_n'(m) = \sum_{m\geq 1} R_n'(m) = -r_n. \qquad \square$$

Using the properties of Euler's gamma function $\Gamma(t)$ (see Section 3.1) we observe the expression

$$F(t) = \left(\frac{\pi}{\sin \pi t}\right)^2 R_n(t) = \frac{\Gamma(n+1-t)^2\Gamma(t)^4}{\Gamma(n+1+t)^2} \tag{7.3}$$

for the integrand in Lemma 7.5.

Lemma 7.6 (Stirling's formula). *In the half-plane* $\operatorname{Re} t > 0$,

$$\ln\Gamma(t) = \left(t - \frac{1}{2}\right)\ln t - t + \ln\sqrt{2\pi} + \rho(t),$$

where the error term $\rho(t)$ *satisfies* $|\rho(t)| \leq c\,(\operatorname{Re} t)^{-1}$ *for some absolute constant* $c > 0$.

As proving this formula is beyond our scope here, we only highlight some underlying ideas behind the proof.

Sketch. For the logarithmic derivative of the gamma function, one can show that

$$\frac{\Gamma'(t)}{\Gamma(t)} = \frac{\mathrm{d}}{\mathrm{d}t} \ln \Gamma(t) = \int_0^\infty \left(\frac{e^{-u}}{u} - \frac{e^{-tu}}{1-e^{-u}} \right) \mathrm{d}u; \tag{7.4}$$

this formula is due to Binet. Integrating this equality intelligently, one gets

$$\ln \Gamma(t) = \left(t - \frac{1}{2}\right) \ln t - t + \ln \sqrt{2\pi} + \int_0^\infty \left(\frac{1}{2} - \frac{1}{u} + \frac{1}{e^u - 1} \right) \frac{e^{-tu}}{u} \,\mathrm{d}u. \tag{7.5}$$

By choosing c to be the maximum of

$$\left| \frac{1}{2} - \frac{1}{u} + \frac{1}{e^u - 1} \right| \frac{1}{u}$$

on the real half-line $u > 0$ (and one can check that the expression is bounded there), we finally find that

$$|\rho(t)| \le c \int_0^\infty |e^{-tu}| \,\mathrm{d}u = c \int_0^\infty e^{-(\operatorname{Re} t)\, u} \,\mathrm{d}u = \frac{c}{\operatorname{Re} t}. \qquad \square$$

Exercise 7.3. (a) Deduce formula (7.5) from Binet's (7.4).
(b) Complete the proof of Lemma 7.6.
(c) Prove Stirling's asymptotic formula for the factorial function

$$n! \sim \sqrt{2\pi n} \left(\frac{n}{e} \right)^n \quad \text{as } n \to \infty,$$

and its corollary

$$\binom{2n}{n} \sim \frac{2^{2n}}{\sqrt{\pi n}} \quad \text{as } n \to \infty$$

for the central binomial coefficients.

Lemma 7.7. *As $n \to \infty$, the following asymptotics is valid: $r_n^{1/n} \to (\sqrt{2} - 1)^4$.*

Proof. In the integral representation of Lemma 7.5 take $C = (n+1)/\sqrt{2}$ and change the variable $t = (n+1)z$. The real parts of $n+1+t$, $n+1-t$ and t are bounded by $c_1 n$ on the contour of integration for some $c_1 > 0$, hence application of Lemma 7.6 to the integrand (7.3) results in

$$\begin{aligned} \ln F(t) &= (2n + 1 - 2t) \ln(n + 1 - t) - 2(n + 1 - t) + (4t - 2) \ln t - 4t \\ &\quad - (2n + 1 + 2t) \ln(n + 1 + t) + 2(n + 1 + t) + 2 \ln(2\pi) + O(n^{-1}) \\ &= 2(n + 1) f(z) + \ln h(z) - 2 \ln(n + 1) + 2 \ln(2\pi) + O(n^{-1}), \end{aligned}$$

where

$$f(z)=(1-z)\ln(1-z)+2z\ln z-(1+z)\ln(1+z), \quad h(z)=\frac{1+z}{z^2(1-z)}$$

and the constant in $O(n^{-1})$ is absolute. This implies that

$$r_n=\frac{2\pi}{ni}\int_{z_0-i\infty}^{z_0+i\infty} e^{2(n+1)f(z)}\frac{1+z}{z^2(1-z)}\left(1+O(n^{-1})\right)\mathrm{d}z$$

for $z_0=1/\sqrt{2}$ and some absolute constant in $O(n^{-1})$.

Consider the function $g(y)=\operatorname{Re}f(z_0+iy)$, that is, the real part of $f(z)$ on the contour of integration. We have

$$\frac{\mathrm{d}}{\mathrm{d}y}g(y)=-\operatorname{Im}\frac{\mathrm{d}f}{\mathrm{d}z}\bigg|_{z=z_0+iy}=\operatorname{Im}\ln(z^{-2}-1),$$

hence $\mathrm{d}g/\mathrm{d}y$ vanishes at $y=0$ only. In a neighbourhood of the point we get $g(y)=g(0)-2^{3/2}y^2+O(y^3)$, so that $g(y)$ has its maximum at $y=0$. Then

$$f(z)=f(z_0)+2^{3/2}(z-z_0)^2+O\big((z-z_0)^3\big)=g(0)+2^{3/2}(z-z_0)^2+O\big((z-z_0)^3\big)$$

on the contour of integration — the maximum of $|e^{f(z)}|$ is attained at $z=z_0$ and it is equal to $e^{f(z_0)}$. Thus, we obtain

$$\lim_{n\to\infty} r_n^{1/(n+1)}=e^{2f(z_0)}=(\sqrt{2}-1)^4,$$

and the result follows. □

Theorem 7.2 (Apéry's theorem). *The number $\zeta(3)$ is irrational.*

Proof. Assume on the contrary that $\zeta(3)$ is rational, $\zeta(3)=a/b$ for some $a,b\in\mathbb{Z}_{>0}$. Since $r_n^{1/n}$ tends to a positive quantity as $n\to\infty$, we conclude that r_n does not vanish for all n sufficiently large. In particular, the *integral* numbers $bd_n^3r_n=ad_n^3q_n-bd_n^3p_n$ are nonzero for all such indices n; this implies that $|bd_n^3r_n|\ge 1$ for all sufficiently large n. On the other hand, $|bd_n^3r_n|^{1/n}\to e^3(\sqrt{2}-1)^4=0.59\ldots<1$, so that $|bd_n^3r_n|<1$ for all n large. The contradiction means that our assumption $\zeta(3)\in\mathbb{Q}$ is invalid. □

7.3 Arithmetic properties of special rational functions

In this part, which is spread over Sections 7.3–7.6, we generalise the construction from Sections 7.1–7.2 to prove the following result.

Theorem 7.3. *At least one of eleven numbers $\zeta(5),\zeta(7),\ldots,\zeta(25)$ is irrational.*

We fix an odd integer $s \ge 7$. Our strategy is constructing *two* sequences of linear forms r_n and $\hat{r}_n$ living in the $\mathbb{Q}$-space $\mathbb{Q} + \mathbb{Q}\zeta(3) + \mathbb{Q}\zeta(5) + \cdots + \mathbb{Q}\zeta(s)$, for which we have a control of the common denominators λ_n of rational coefficients and an elementary access to their asymptotic behaviour as $n \to \infty$; more importantly, the two coefficients of $\zeta(3)$ in these forms are proportional (with factor 7), so that $7r_n - \hat{r}_n$ belongs to the space $\mathbb{Q} + \mathbb{Q}\zeta(5) + \cdots + \mathbb{Q}\zeta(s)$. Finally, using $7r_n - \hat{r}_n > 0$ and the asymptotics $\lambda_n(7r_n - \hat{r}_n) \to 0$ as $n \to \infty$ of the linear forms

$$\lambda_n(7r_n - \hat{r}_n) \in \mathbb{Z} + \mathbb{Z}\zeta(5) + \mathbb{Z}\zeta(7) + \cdots + \mathbb{Z}\zeta(s)$$

when $s = 25$, we conclude that it cannot happen that all the quantities $\zeta(5), \zeta(7), \ldots, \zeta(25)$ are rational.

More precisely, our linear forms assume the form

$$r_n = \sum_{\nu=1}^{\infty} R_n(\nu) \quad \text{and} \quad \hat{r}_n = \sum_{\nu=1}^{\infty} R_n(\nu - \tfrac{1}{2}), \tag{7.6}$$

where the rational-function summand $R_n(t)$ is defined as follows:

$$\begin{aligned} R_n(t) &= \frac{n!^{s-5} \prod_{j=1}^{n}(t-j) \cdot \prod_{j=1}^{n}(t+n+j) \cdot 2^{6n} \prod_{j=1}^{3n}(t-n-\frac{1}{2}+j)}{\prod_{j=0}^{n}(t+j)^s} \\ &= \frac{2^{6n} n!^{s-5} \prod_{j=0}^{6n}(t-n+\frac{1}{2}j)}{\prod_{j=0}^{n}(t+j)^{s+1}}. \end{aligned} \tag{7.7}$$

We first discuss a general rational function $S(t)$ of the form

$$S(t) = \frac{P(t)}{(t-t_1)^{s_1}(t-t_2)^{s_2} \cdots (t-t_q)^{s_q}},$$

whose denominator has degree larger than its numerator, so that its unique partial-fraction decomposition assumes the form

$$S(t) = \sum_{j=1}^{q} \sum_{i=1}^{s_j} \frac{b_{i,j}}{(t-t_j)^i}.$$

The coefficients here can be computed on the basis of explicit formula

$$b_{i,j} = \frac{1}{(s_j - i)!} \left(S(t)(t-t_j)^{s_j} \right)^{(s_j - i)} \Big|_{t=t_j}$$

for all i, j in question. (This procedure is seen in action in the examples discussed in Lemma 7.1 and Exercise 7.1, when all the exponents s_j are equal to 1.) It also means that the function $R(t)$ in (7.7) can be written as

$$R(t) = \sum_{i=1}^{s} \sum_{k=0}^{n} \frac{a_{i,k}}{(t+k)^i} \tag{7.8}$$

with the recipe to compute the coefficients $a_{i,k}$ in its partial-fraction decomposition. At the same time, the function $R(t)$ is a product of 'simpler' rational functions given Lemma 7.1 and Exercise 7.1, with all coefficients of their partial fractions being integral.

Lemma 7.8. *Let $k_1, \dots, k_q$ be pairwise distinct numbers from the set $\{0, 1, \dots, n\}$ and $s_1, \dots, s_q$ positive integers. Then the coefficients in the expansion*

$$\frac{1}{\prod_{j=1}^{q}(t+k_j)^{s_j}} = \sum_{j=1}^{q}\sum_{i=1}^{s_j}\frac{b_{i,j}}{(t+k_j)^i}$$

satisfy

$$d_n^{s-i} b_{i,j} \in \mathbb{Z}, \quad \text{where } i = 1, \dots, s_j \text{ and } j = 1, \dots, q, \tag{7.9}$$

where $s = s_1 + \cdots + s_q$.

In particular,

$$d_n^{s-i} a_{i,k} \in \mathbb{Z}, \quad \text{where } i = 1, \dots, s \text{ and } k = 0, 1, \dots, n, \tag{7.10}$$

for the coefficients in (7.8).

Proof. Denote the rational function in question by $S(t)$. The statement is trivially true when $q = 1$, therefore we assume that $q \ge 2$. In view of the symmetry of the data, it is sufficient to demonstrate the inclusions (7.9) for $j = 1$. Differentiating a related product m times, for any $m \ge 0$, we obtain

$$\begin{aligned} \frac{1}{m!}\big(S(t)(t+k_1)^{s_1}\big)^{(m)} &= \frac{1}{m!}\bigg(\prod_{j=2}^{q}(t+k_j)^{-s_j}\bigg)^{(m)} \\ &= \sum_{\substack{\ell_2,\dots,\ell_q \ge 0 \\ \ell_2+\cdots+\ell_q = m}} \prod_{j=2}^{q}\frac{1}{\ell_j!}\big((t+k_j)^{-s_j}\big)^{(\ell_j)} \\ &= \sum_{\substack{\ell_2,\dots,\ell_q \ge 0 \\ \ell_2+\cdots+\ell_q = m}} \prod_{j=2}^{q}(-1)^{\ell_j}\binom{s_j+\ell_j-1}{\ell_j}(t+k_j)^{-(s_j+\ell_j)}. \end{aligned}$$

This implies that

$$b_{i,1} = \sum_{\substack{\ell_2,\dots,\ell_q \ge 0 \\ \ell_2+\cdots+\ell_q = s_1-i}} \prod_{j=2}^{q}(-1)^{\ell_j}\binom{s_j+\ell_j-1}{\ell_j}\frac{1}{(k_j-k_1)^{s_j+\ell_j}}$$

for $i = 1, \dots, s_1$. Using $d_n/(k_j - k_1) \in \mathbb{Z}$ for $j = 2, \dots, q$ and $\sum_{j=2}^{q}(s_j + \ell_j) = s - i$ for each individual summand, we deduce the desired inclusion in (7.9) for $j = 1$, hence for any j.

The second claim in the lemma follows from considering $R(t)$ as a product of the 'simpler' rational functions from Lemma 7.1 and Exercise 7.1. $\square$

Lemma 7.9. *For the coefficients $a_{i,k}$ in (7.8), we have*

$$a_{i,k} = (-1)^{i-1} a_{i,n-k} \quad \textit{for } k = 0, 1, \dots, n \textit{ and } i = 1, \dots, s,$$

so that

$$\sum_{k=0}^{n} a_{i,k} = 0 \quad \textit{for } i \textit{ even.}$$

Proof. Since s is odd, the function (7.7) possesses the following (well-poised) symmetry: $R(-t-n) = -R(t)$. Substitution of the relation into (7.8) results in

$$\begin{aligned} -\sum_{i=1}^{s} \sum_{k=0}^{n} \frac{a_{i,k}}{(t+k)^i} &= \sum_{i=1}^{s} \sum_{k=0}^{n} \frac{a_{i,k}}{(-t-n+k)^i} = \sum_{i=1}^{s} (-1)^i \sum_{k=0}^{n} \frac{a_{i,k}}{(t+n-k)^i} \\ &= \sum_{i=1}^{s} (-1)^i \sum_{k=0}^{n} \frac{a_{i,n-k}}{(t+k)^i}, \end{aligned}$$

and the identities in the lemma follow from the uniqueness of decomposition into partial fractions. The second statement follows from

$$\sum_{k=0}^{n} a_{i,k} = (-1)^{i-1} \sum_{k=0}^{n} a_{i,n-k} = (-1)^{i-1} \sum_{k=0}^{n} a_{i,k}. \qquad \square$$

7.4 Arithmetic properties of linear forms in zeta values

We now take a closer look at the quantities defined in (7.6).

Lemma 7.10. *For each n,*

$$r_n = \sum_{\substack{i=2 \\ i \text{ odd}}}^{s} a_i \zeta(i) + a_0 \quad \textit{and} \quad \hat{r}_n = \sum_{\substack{i=2 \\ i \text{ odd}}}^{s} a_i (2^i - 1)\zeta(i) + \hat{a}_0,$$

with the following inclusions available:

$$d_n^{s-i} a_i \in \mathbb{Z} \quad \textit{for } i = 3, 5, \dots, s, \quad \textit{and} \quad d_n^s a_0,\ d_n^s \hat{a}_0 \in \mathbb{Z}.$$

Notice that

$$(2^i - 1)\zeta(i) = \sum_{\ell=1}^{\infty} \frac{1}{(\ell - \frac{1}{2})^i}$$

for $i \geq 2$.

Proof. Our strategy here is to write the series in (7.6) using the partial-fraction decomposition (7.8) of $R(t)$. To treat the first sum r_n we additionally introduce an auxiliary parameter $z > 0$, which we later specialise to $z = 1$:

$$\begin{aligned} r_n(z) &= \sum_{\nu=1}^{\infty} R_n(\nu) z^\nu = \sum_{\nu=1}^{\infty} \sum_{i=1}^{s} \sum_{k=0}^{n} \frac{a_{i,k} z^\nu}{(\nu+k)^i} \\ &= \sum_{i=1}^{s} \sum_{k=0}^{n} a_{i,k} z^{-k} \sum_{\nu=1}^{\infty} \frac{z^{\nu+k}}{(\nu+k)^i} = \sum_{i=1}^{s} \sum_{k=0}^{n} a_{i,k} z^{-k} \biggl(\mathrm{Li}_i(z) - \sum_{\ell=1}^{k} \frac{z^\ell}{\ell^i} \biggr) \\ &= \sum_{i=1}^{s} \mathrm{Li}_i(z) \sum_{k=0}^{n} a_{i,k} z^{-k} - \sum_{i=1}^{s} \sum_{k=0}^{n} \sum_{\ell=1}^{k} \frac{a_{i,k} z^{-(k-\ell)}}{\ell^i}, \end{aligned}$$

where

$$\mathrm{Li}_i(z) = \sum_{\ell=1}^{\infty} \frac{z^\ell}{\ell^i}$$

for $i = 1, \dots, s$ are the polylogarithmic functions. The latter are well defined at $z = 1$ for $i \ge 2$, where $\mathrm{Li}_i(1) = \zeta(i)$, while $\mathrm{Li}_1(z) = -\log(1-z)$ does not have a limit as $z \to 1^-$. By taking the limit as $z \to 1^-$ in the above derivation and using $R_n(\nu) = O(\nu^{-2})$ as $\nu \to \infty$, we conclude that

$$\sum_{k=0}^{n} a_{1,k} = \lim_{z \to 1^-} \sum_{k=0}^{n} a_{1,k} z^{-k} = 0,$$

and

$$r_n = \sum_{i=2}^{s} \zeta(i) \sum_{k=0}^{n} a_{i,k} - \sum_{i=1}^{s} \sum_{k=0}^{n} a_{i,k} \sum_{\ell=1}^{k} \frac{1}{\ell^i}. \tag{7.11}$$

We proceed similarly for $\hat{r}_n$, omitting introduction of the auxiliary parameter z. Since $R(t)$ in (7.7) vanishes at $t = -\frac{1}{2}, -\frac{3}{2}, \dots, -n + \frac{1}{2}$, we can shift the starting point of summation for $\hat{r}_n$ to $t = -m - \frac{1}{2}$, where $m = \lfloor \frac{n-1}{2} \rfloor$, so that

$$\begin{aligned} \hat{r}_n &= \sum_{\nu=-m}^{\infty} R_n(\nu - \tfrac{1}{2}) = \sum_{\nu=-m}^{\infty} \sum_{i=1}^{s} \sum_{k=0}^{n} \frac{a_{i,k}}{(\nu + k - \frac{1}{2})^i} \\ &= \sum_{i=1}^{s} \sum_{k=0}^{n} a_{i,k} \sum_{\nu=-m}^{\infty} \frac{1}{(\nu + k - \frac{1}{2})^i} \\ &= \sum_{i=1}^{s} \sum_{k=0}^{m} a_{i,k} \sum_{\nu=-m}^{\infty} \frac{1}{(\nu + k - \frac{1}{2})^i} + \sum_{i=1}^{s} \sum_{k=m+1}^{n} a_{i,k} \sum_{\nu=-m}^{\infty} \frac{1}{(\nu + k - \frac{1}{2})^i} \end{aligned}$$

$$
\begin{aligned}
&= \sum_{i=1}^{s}\sum_{k=0}^{m} a_{i,k}\biggl(\sum_{\ell=k-m}^{0}\frac{1}{(\ell-\frac12)^i}+\sum_{\ell=1}^{\infty}\frac{1}{(\ell-\frac12)^i}\biggr)\\
&\quad+\sum_{i=1}^{s}\sum_{k=m+1}^{n} a_{i,k}\biggl(\sum_{\ell=1}^{\infty}\frac{1}{(\ell-\frac12)^i}-\sum_{\ell=1}^{k-m-1}\frac{1}{(\ell-\frac12)^i}\biggr)\\
&= \sum_{i=2}^{s}(2^i-1)\zeta(i)\sum_{k=0}^{n}a_{i,k}+\sum_{i=1}^{s}\sum_{k=0}^{m}a_{i,k}\sum_{\ell=0}^{m-k}\frac{(-1)^i}{(\ell+\frac12)^i}\\
&\quad-\sum_{i=1}^{s}\sum_{k=m+1}^{n}a_{i,k}\sum_{\ell=1}^{k-m-1}\frac{1}{(\ell-\frac12)^i}.
\end{aligned}
\tag{7.12}
$$

Now the statement of the lemma follows from the representations in (7.11) and (7.12), Lemma 7.9, the inclusions (7.10) of Lemma 7.8 and

$$
\begin{aligned}
d_n^i\sum_{\ell=1}^{k}\frac{1}{\ell^i} &\in \mathbb{Z} \quad \text{for } 0\le k\le n \text{ and } i\ge 1,\\
d_n^i\sum_{\ell=0}^{m-k}\frac{(-1)^i}{(\ell+\frac12)^i} &\in \mathbb{Z} \quad \text{for } 0\le k\le m \text{ and } i\ge 1,\\
d_{n-1}^i\sum_{\ell=1}^{k-m-1}\frac{1}{(\ell-\frac12)^i} &\in \mathbb{Z} \quad \text{for } m+1\le k\le n \text{ and } i\ge 1. \qquad \square
\end{aligned}
$$

7.5 Asymptotic behaviour

In this section we make frequent use of Stirling's asymptotic formula for the factorial function from Exercise 7.3(c).

Because the rational function $R_n(t)$ in (7.7) vanishes at $1,2,\dots,n$ and at $\frac12,\frac32,\dots,n-\frac12$, the sums (7.6) can be alternatively written as

$$
r_n=\sum_{\nu=n+1}^{\infty}R_n(\nu)=\sum_{k=0}^{\infty}c_k \quad\text{and}\quad \hat r_n=\sum_{\nu=n+1}^{\infty}R_n(\nu-\tfrac12)=\sum_{k=0}^{\infty}\hat c_k,
$$

with the involved summands

$$
\begin{aligned}
c_k=R_n(n+1+k)&=\frac{2^{6n}n!^{s-5}\prod_{j=0}^{6n}(k+1+\frac12 j)}{\prod_{j=0}^{n}(n+k+1+j)^{s+1}}\\
&=\frac{n!^{s-5}(6n+2k+2)!\,(n+k)!^{s+1}}{2\,(2k+1)!\,(2n+k+1)!^{s+1}}
\end{aligned}
\tag{7.13}
$$

and

$$\hat{c}_k = R_n(n + \tfrac{1}{2} + k) = \frac{2^{6n} n!^{s-5} \prod_{j=0}^{6n}(k + \frac{1}{2} + \frac{1}{2}j)}{\prod_{j=0}^{n}(n + k + \frac{1}{2} + j)^{s+1}}$$

strictly *positive*. Observe that

$$\begin{aligned}
\frac{c_k}{\hat{c}_k} &= \frac{\prod_{j=0}^{6n}(2k+2+j)}{\prod_{j=0}^{6n}(2k+1+j)} \cdot \left(\prod_{j=0}^{n} \frac{n+k+\frac{1}{2}+j}{n+k+1+j}\right)^{s+1} \\
&= \frac{6n+2k+2}{2k+1} \cdot \left(2^{-2(n+1)} \frac{\binom{4n+2k+2}{2n+k+1}}{\binom{2n+2k}{n+k}}\right)^{s+1} \\
&\sim \frac{6n+2k+2}{2k+1} \left(\frac{n+k}{2n+k+1}\right)^{(s+1)/2} \qquad \text{as } n+k \to \infty. \qquad (7.14)
\end{aligned}$$

Lemma 7.11. *For $s \ge 7$ odd,*

$$\lim_{n\to\infty} r_n^{1/n} = \lim_{n\to\infty} \hat{r}_n^{1/n} = g(x_0) \quad \text{and} \quad \lim_{n\to\infty} \frac{r_n}{\hat{r}_n} = 1$$

where

$$g(x) = \frac{2^6 (x+3)^6 (x+1)^{s+1}}{(x+2)^{2(s+1)}}$$

and x_0 is the unique positive zero of the polynomial

$$x(x+2)^{(s+1)/2} - (x+3)(x+1)^{(s+1)/2}.$$

Proof. We have

$$\frac{c_{k+1}}{c_k} = \frac{(k+3n+\frac{3}{2})(k+3n+2)}{(k+1)(k+\frac{3}{2})} \left(\frac{k+n+1}{k+2n+2}\right)^{s+1} \sim f\left(\frac{k}{n}\right)^2 \qquad (7.15)$$

as $n + k \to \infty$, where

$$f(x) = \frac{x+3}{x} \left(\frac{x+1}{x+2}\right)^{(s+1)/2}.$$

For an ease of notation write $q = (s+1)/2 \ge 4$. Since

$$\frac{f'(x)}{f(x)} = \frac{1}{x+3} - \frac{1}{x} + q\left(\frac{1}{x+1} - \frac{1}{x+2}\right) = \frac{(q-3)x^2 + 3(q-3)x - 6}{x(x+1)(x+2)(x+3)}$$

and the quadratic polynomial in the latter numerator has a unique positive zero x_1, the function $f(x)$ monotone decreases from $+\infty$ to $f(x_1)$ when x ranges from 0 to x_1 and then monotone increases from $f(x_1)$ to $f(+\infty) = 1$ (not attaining the value!) when x ranges from x_1 to $+\infty$. In particular, there is exactly one positive solution x_0 to $f(x) = 1$. Notice that $0 < x_0 < 1$, because $f(1) = 4 \cdot (2/3)^q < 1$.

The information gained and asymptotics in (7.15) imply that $c_{k+1}/c_k > 1$ for the indices $k < x_0 n - \gamma\sqrt{n}$ and $c_{k+1}/c_k < 1$ for $k > x_0 n + \gamma\sqrt{n}$ for an appropriate choice of $\gamma > 0$ dictated by application of Stirling's formula to the factorials defining c_k in (7.13). This means that the asymptotic behaviour of the sum $r_n = \sum_{k=0}^{\infty} c_k$ is determined by the asymptotics of c_{k_0} and its neighbours c_k, where $k_0 = k_0(n) \sim x_0 n$ and $|k - k_0| \le \gamma\sqrt{n}$, so that

$$
\begin{aligned}
\lim_{n\to\infty} r_n^{1/n} &= \lim_{n\to\infty} c_{k_0(n)}^{1/n} \\
&= \lim_{n\to\infty} \biggl(\Bigl(\frac{n}{e}\Bigr)^{(s-5)n} \Bigl(\frac{6n+2k_0+2}{e}\Bigr)^{6n+2k_0+2} \Bigl(\frac{e}{2k_0+1}\Bigr)^{2k_0+1} \\
&\qquad \times \Bigl(\frac{n+k_0}{e}\Bigr)^{(s+1)(n+k_0)} \Bigl(\frac{e}{2n+k_0+1}\Bigr)^{(s+1)(2n+k_0+1)} \biggr)^{1/n} \\
&= \frac{(2x_0+6)^{2x_0+6}(x_0+1)^{(s+1)(x_0+1)}}{(2x_0)^{2x_0}(x_0+2)^{(s+1)(x_0+2)}} \\
&= \frac{2^6(x_0+3)^6(x_0+1)^{s+1}}{(x_0+2)^{2(s+1)}} \cdot f(x_0)^{2x_0} = g(x_0).
\end{aligned}
$$

It now follows from (7.14) that

$$
\frac{\hat{c}_{k+1}}{\hat{c}_k} \sim \frac{c_{k+1}}{c_k} \quad \text{as } n+k \to \infty, \tag{7.16}
$$

so that the above analysis applies to the sum $\hat{r}_n = \sum_{k=0}^{\infty} \hat{c}_k$ as well, and its asymptotic behaviour is determined by the asymptotics of $\hat{c}_{k_0}$ and its neighbours $\hat{c}_k$, where $k_0 = k_0(n) \sim x_0 n$ and $|k - k_0| \le \hat{\gamma}\sqrt{n}$. From (7.16) we deduce that the limits of $\hat{c}_{k_0(n)}^{1/n}$ and $c_{k_0(n)}^{1/n}$ as $n \to \infty$ coincide, hence $\hat{r}_n^{1/n} \to g(x_0)$ as $n \to \infty$. In addition to this, we also get

$$
\lim_{n\to\infty} \frac{r_n}{\hat{r}_n} = \lim_{n\to\infty} \frac{c_{k_0(n)}}{\hat{c}_{k_0(n)}} = \lim_{n\to\infty} \frac{6n+2k_0+2}{2k_0+1} \Bigl(\frac{n+k_0}{2n+k_0+1}\Bigr)^{(s+1)/2} = f(x_0),
$$

which leads to the remaining limiting relation. □

7.6 One of the numbers $\zeta(5), \zeta(7), \ldots, \zeta(25)$ is irrational

We now combine the information gathered about the linear forms r_n and $\hat{r}_n$ to conclude our proof of Theorem 7.3.

We choose $s = 25$ and apply Lemma 7.11 to find out that $7r_n - \hat{r}_n > 0$ for n sufficiently large, and

$$\lim_{n\to\infty} (7r_n - \hat{r}_n)^{1/n} = g(x_0) = \exp(-25.292363\dots),$$

where $x_0 = 0.00036713\dots$ is the positive zero of $x(x+2)^{13} - (x+3)(x+1)^{13}$. Assuming that the odd zeta values from $\zeta(5)$ to $\zeta(25)$ are all rational and denoting by b their common denominator, we use Lemmas 7.2 and 7.10 to conclude that the sequence of *positive integers*

$$bd_n^{25}(7r_n - \hat{r}_n)$$

tends to 0 as $n \to \infty$; contradiction. Thus, at least one of the numbers $\zeta(5), \zeta(7), \dots, \zeta(25)$ is irrational.

Chapter notes

Numerous proofs of Apéry's theorem (Theorem 7.2) are now recorded; in our exposition we follow closely the version given in [58]. The story behind the original proof of Apéry [3] (together with the completed proof!) is beautifully presented in [64]. A proof given shortly after by F. Beukers [12] uses real-valued triple integrals; it is still considered as most elegant and sources further research [15, 16, 66] in the irrationality direction. A 2004 historical account of the development around Apéry's and Rivoal's theorem can be found in [30].

The proof of Lemma 7.11 (elementary asymptotics of linear forms) is inspired by the methodology and examples from de Bruijn's book [17], which is a definite recommendation for learning techniques in asymptotics analysis.

The trick, used in Theorem 7.3, of eliminating an 'unwanted' term of $\zeta(3)$ in linear forms in odd zeta values finds further applications. One of the most recent news in this direction is the result of L. Lai and P. Yu [51] that at least $\frac{1}{10}\sqrt{s/\ln s}$ numbers on the list $\zeta(3), \zeta(5), \dots, \zeta(s)$ are irrational, where $s > 10^4$ is odd; this in turn builds on the earlier work [31].

Chapter 8

Hilbert's seventh problem

The following problem posed by Hilbert in 1900 was resolved in the 1930s independently by A. Gelfond and Th. Schneider.

Theorem 8.1 (Hilbert's seventh problem, Gelfond–Schneider theorem). *Let α and β be algebraic, $\alpha \neq 0, 1$ and β irrational. Then α^β is transcendental.*

In this chapter we expose two different proofs of Theorem 8.1. Our first proof uses the so-called interpolation determinants of M. Laurent, while the second one (only sketched here) is the original proof of Schneider. In both proofs, the constructions depend on a sufficiently large natural parameter N.

Exercise 8.1. Show that the statement of Theorem 8.1 is equivalent to the following: If α_1, α_2 are nonzero algebraic numbers such that the quotient

$$\gamma = \frac{\ln \alpha_1}{\ln \alpha_2}$$

is irrational, then γ is transcendental.

This is the form, in which Theorem 8.1 was proven by Gelfond.

8.1 Reduction of proof

Inspired by Exercise 8.1, from now on we use the notation $\alpha_1 = \alpha$ and $\alpha_2 = \alpha^\beta$. Define the (non-square!) matrix

$$\mathcal{M} = \|a_{r,s}^{u,v}\|, \quad a_{r,s}^{u,v} = (r + s\beta)^u (\alpha_1^r \alpha_2^s)^v,$$

where $0 \le r, s < 2N$ and $0 \le u < K = \lfloor N \ln N \rfloor$, $0 \le v < L = \left\lfloor \dfrac{N}{\ln N} \right\rfloor$, whose columns are indexed by pairs u, v (the total number of which is $KL \sim N^2$), while the rows are indexed by pairs r, s (and there are exactly $4N^2$ such pairs).

Lemma 8.1. *The rank of the matrix $\mathcal{M}$ is equal to KL (that is, maximal possible).*

The lemma implies the existence of a nonzero minor of $\mathcal{M}$ of maximal order. Choose and fix one of these nonzero minors, say Δ, and denote $\mathcal{L}$ the corresponding collection of rows r, s:

$$\Delta = \det \|a_{r,s}^{u,v}\|_{(r,s)\in\mathcal{L}}^{0\le u<K,\,0\le v<L} \neq 0.$$

Lemma 8.2. *Eventually, the estimate* $\ln|\Delta| \le -N^4$ *holds.*

Lemma 8.3. *If $\alpha, \beta, \alpha^\beta$ are algebraic numbers, then* $\ln|\Delta| \ge -\frac{1}{2}N^4$ *for all sufficiently large N.*

Proof of Theorem 8.1. Assuming the three numbers $\alpha, \beta, \alpha^\beta$ are algebraic, we find the the estimates in Lemmas 8.2 and 8.3 contradictory. This shows the truth of Theorem 8.1. □

In order to move further, we will introduce some more notation. For an algebraic number α, denote by $\mathbb{Q}(\alpha)$ the algebraic extension of the field of rationals that contains all polynomials (and rational functions!) of α with rational coefficients. The notation $[\mathbb{Q}(\alpha) : \mathbb{Q}]$ is then used to denote the degree of algebraic α. If an algebraic field K (that is, a field whose all elements are algebraic numbers) can be generated by finitely many algebraic numbers $\alpha_1, \dots, \alpha_m$, then there is also a single generator α of it called a *primitive element*, $K = \mathbb{Q}(\alpha)$; then $[K : \mathbb{Q}] = [\mathbb{Q}(\alpha) : \mathbb{Q}]$. In a similar way, the intermediate degrees $[K : \mathbb{Q}(\alpha)]$ are introduced, when K is an algebraic extension of $\mathbb{Q}(\alpha)$.

If $P(x_1, \dots, x_m)$ is a polynomial, then the maximum of the absolute values of its coefficients is called the *height* and denoted $H(P)$, while the sum of the absolute values of its coefficients is called the *length* and denoted $L(P)$. The height and length of an algebraic α corresponds to the related characteristics of the minimal primitive polynomial for α.

Theorem 8.2 (Liouville-type theorem; compare with Exercise 6.7). *Let $\alpha_1, \dots, \alpha_m$ be algebraic, $K = \mathbb{Q}(\alpha_1, \dots, \alpha_m)$, and let $P(x_1, \dots, x_m)$ be a polynomial with integral coefficients. Then either $P(\alpha_1, \dots, \alpha_m) = 0$ or*

$$|P(\alpha_1, \dots, \alpha_m)| \ge L(P)^{-[K:\mathbb{Q}]} \cdot \prod_{i=1}^{m} L(\alpha_i)^{-\deg_{x_i} P \cdot [K:\mathbb{Q}(\alpha_i)]}.$$

Proof of Lemma 8.3. Consider the polynomial

$$P(x_1, x_2, x_3) = \det \|(r + sx_3)^u (x_1^r x_2^s)^v\|_{(r,s)\in\mathcal{L}}^{0\le u<K,\, 0\le v<L} \in \mathbb{Z}[x_1, x_2, x_3].$$

Note that $P \not\equiv 0$ since $P(\alpha_1, \alpha_2, \beta) = \Delta \neq 0$ (by Lemma 8.1 and our choice of Δ).

To apply Theorem 8.2, notice that in our case the numbers $\alpha_1, \alpha_2, \alpha_3 = \beta$ are fixed, so that the degrees of algebraic extensions $[K : \mathbb{Q}]$ and $[K : \mathbb{Q}(\alpha_i)]$ are fixed as well. Furthermore,

$$\begin{aligned}
\deg_{x_1} P &\le (2N \cdot L) \cdot KL \le \frac{2N^4}{\ln N}, \\
\deg_{x_2} P &\le (2N \cdot L) \cdot KL \le \frac{2N^4}{\ln N}, \\
\deg_{x_3} P &\le K \cdot KL \le N^3 \ln N
\end{aligned}$$

and

$$L(P) \le (KL)! \cdot (4N)^{K\cdot KL} \le e^{N^2 \ln(N^2)} \cdot e^{N^3 \ln N \cdot \ln(4N)} \le e^{2N^3 \ln^2 N}$$

for all sufficiently large N. (The multiple $(KL)!$ corresponds to the numbers of terms in expanding the determinant, and $(4N)^{K\cdot KL}$ estimates the length of each of these terms from above.) By Theorem 8.2 we obtain

$$\begin{aligned}
\ln|\Delta| = \ln|P(\alpha_1, \alpha_2, \beta)| &\ge -c\left(N^3 \ln^2 N + 2 \cdot \frac{2N^4}{\ln N} + N^3 \ln N\right) \\
&\ge -c_1 \frac{N^4}{\ln N} \ge -\frac{1}{2} N^4
\end{aligned}$$

for all sufficiently large N, which completes the proof of Lemma 8.3. □

8.2 Interpolation determinants

Lemma 8.4. *Let $R > \rho > 0$. Assume we have M complex numbers $\xi_1, \dots, \xi_M$ inside the disc $|\xi| \le \rho$ and M functions $f_1(z), \dots, f_M(z)$ which are analytic in the disc $|z| \le R$ such that*

$$|f_j|_R = \max_{|z|\le R} |f_j(z)| \le S, \quad j = 1, \dots, M.$$

Denote $\delta = \det \|f_i(\xi_j)\|_{1\le i,j\le M}$. Then the following estimate holds:

$$|\delta| \le M! \left(\frac{R}{\rho}\right)^{-M(M-1)/2} S^M.$$

Remark. The determinant δ is called an *interpolation determinant.* The reason behind this definition is as follows.

Given a function $g(z)$ and numbers $\xi_1, \dots, \xi_M$, a standard interpolation problem is finding a polynomial $F(z)$ of degree at most $M-1$ such that $F(\xi_i) = g(\xi_i)$ for all $i = 1, \dots, M$. Let complexify the problem by introducing M analytic functions $f_1(z), \dots, f_M(z)$ and asking to determine the coefficients in representation $F(z) = a_1 f_1(z) + \dots + a_M f_M(z)$ such that $F(\xi_i) = g(\xi_i)$ for all $i = 1, \dots, M$. (The particular choice $f_j(z) = z^{j-1}$ corresponds to the standard interpolation problem.) In order to solve the corresponding linear system

$$a_1 f_1(\xi_i) + \dots + a_M f_M(\xi_i) = g(\xi_i), \quad i = 1, \dots, M,$$

we need the nonvanishing of its determinant, which is exactly equal to δ. (Of course, the system has a unique up to a scalar solution iff $\delta \neq 0$, which can be found by using, for example, Cramer's rule.)

Proof of Lemma 8.2. Our choice of the functions and points is:

$$f_{u,v}(z) = z^u e^{vz \ln \alpha}, \ 0 \le u < K, \ 0 \le v < L, \quad \xi_{r,s} = r + s\beta, \ (r,s) \in \mathcal{L},$$

where $\ln \alpha$ is a fixed branch of the logarithmic function. Note that $M = KL \sim N^2$. Since

$$|\xi_{r,s}| = |r + s\beta| \le 2N(1 + |\beta|),$$

take $\rho = 2N(1+|\beta|)$ and $R = e^5 \rho$. Furthermore, eventually we have in the disc $|z| \le R$

$$|f_{u,v}(z)| = |z^u e^{vz \ln \alpha}| \le R^K \cdot e^{RL|\ln \alpha|} \le \exp\left(c \frac{N^2}{\ln N}\right) \le e^{N^2},$$

so that Lemma 8.4 is applicable with $S = e^{N^2}$:

$$|\delta| = |\Delta| \le M! e^{-5M(M-1)/2} \cdot (e^{N^2})^{N^2} \le (N^2)^{N^2} e^{-5N^4/2} e^{N^4} \le e^{-N^4}$$

for all sufficiently large N. This proves Lemma 8.2. □

Proof of Lemma 8.4. Consider the auxiliary function

$$F(z) = \det \|f_i(\xi_j z)\|_{1 \le i,j \le M},$$

so that $F(1) = \delta$. We shall demonstrate that $\operatorname{ord}_{z=0} F(z) \ge M(M-1)/2$.

Note that $F(z)$ depends on each function $f_i(z)$ linearly, that is, if $f_i = C_1 f_i^{(1)} + C_2 f_i^{(2)}$, then the determinant $F(z)$ equals the sum of the corresponding determinants $F^{(1)}$ and $F^{(2)}$, multiplied by C_1 and C_2, respectively ($F^{(j)}$ is obtained from F by putting $f_i^{(j)}$ instead of f_i on the ith column).

Writing each of the functions $f_i(z)$ in the form $\widetilde{f}_i(z) + z^{M(M-1)/2} g_i(z)$, where $\widetilde{f}_i(z)$ is a polynomial of degree at most $M(M-1)/2 - 1$ and $g_i(z)$ is analytic at the origin, it is therefore sufficient to verify the estimate $\operatorname{ord}_{z=0} F(z) \ge M(M-1)/2$ for the determinant

$$\widetilde{F}(z) = \det \|\widetilde{f}_i(\xi_j z)\|_{1\le i,j\le M}$$

instead. Again, the linearity and expression of each polynomial $\widetilde{f}_i(z)$ as a (finite) sum of monomials Cz^l, allows us to reduce the verification to the particular case $f_i(z) = z^{l_i}$ for $i = 1, \dots, M$. Then

$$F(z) = \det \|\xi_j^{l_i} z^{l_i}\|_{1\le i,j\le M} = z^{l_1+\dots+l_M} \cdot \det \|\xi_j^{l_i}\|_{1\le i,j\le M}.$$

If $l_{i_1} = l_{i_2}$ for some $i_1 \ne i_2$, then the latter determinant involves equal rows (of indices i_1, i_2), so that $F(z) \equiv 0$ and $\operatorname{ord}_{z=0} F(z) \ge M(M-1)/2$ is automatically satisfied. Otherwise, all l_i are pairwise distinct, and the latter representation implies

$$\operatorname{ord}_{z=0} F(z) = l_1 + \dots + l_M \ge 0 + 1 + 2 + \dots + (M-1) = \frac{M(M-1)}{2},$$

which is the required bound.

Thus, we have shown that the function $G(z) = F(z) \cdot z^{-M(M-1)/2}$ is analytic in the disc $|z| \le R$. In addition, $\delta = F(1) = G(1)$, so that by the maximum modulus principle

$$\begin{aligned}|\delta| = |G(1)| \le |G(z)|_{R/\rho} &= \left(\frac{R}{\rho}\right)^{-M(M-1)/2} \cdot |F(z)|_{R/\rho} \\ &\le \left(\frac{R}{\rho}\right)^{-M(M-1)/2} \cdot M! S^M.\end{aligned}$$

□

Lemma 8.4 is basically a generalization of the following classical result.

Lemma 8.5 (Schwarz lemma). *Let $f(z)$ be a holomorphic map of the disc $|z| \le 1$ onto itself such that $|f(z)| \le 1$ and $f(0) = 0$. Then $|f(z)| \le |z|$.*

Proof. For the holomorphic in the unit disc function $g(z) = f(z)/z$, the maximum modulus principle implies

$$|g(z)| \le |g(z)|_1 = |f(z)|_1 \le 1,$$

which leads to the required result. □

8.3 Rank of interpolation matrix

Proof of Lemma 8.1. Assume on the contrary that the rank of $\mathcal{M}$ is strictly less then KL, so that the columns of the matrix are linearly dependent. The latter means that there exists a polynomial

$$P(x,y)=\sum_{u=0}^{K-1}\sum_{v=0}^{L-1}C_{uv}x^u y^v \not\equiv 0$$

such that

$$P(r+s\beta,\alpha_1^r\alpha_2^s)=0 \quad \text{for all } 0\le r<2N,\ 0\le s<2N. \tag{8.1}$$

Our nearest goal is to show that the equality in (8.1) is violated for at least one pair r, s.

Remark. In general, the set of zeros of a generic 2-variable polynomial $P(x,y)$ is infinite and forms a 1-dimensional variety in $\mathbb{C}^2$. However, in our situation conditions (8.1) mean that the polynomial $P(x,y)$ has 'too many' zeros along the group $G\simeq\mathbb{C}_+\times\mathbb{C}_\times$ in $\mathbb{C}^2$ equipped with group operation $(m_1,n_1)\cdot(m_2,n_2)=(m_1+m_2,n_1n_2)$ and generators $(1,\alpha)$ and (β,α^β). It is not hard to see that

$$(1,\alpha)^r\cdot(\beta,\alpha^\beta)^s=(r+s\beta,\alpha_1^r\alpha_2^s) \quad \text{for all } r,s\in\mathbb{Z}.$$

This interpretation motivates considering a more general problem of estimating the number of zeros of a polynomial $P(x_1,\dots,x_m)$ on the set which possesses a group structure in $\mathbb{C}^m$. There are numerous results in this direction in the last five decades, including famous Baker's linear forms in logarithms [7, 19, 78].

Lemma 8.6. *Let $P\in\mathbb{C}[x,y]$, $P\not\equiv 0$, $\deg_x P<K$ and $\deg_y P<L$. Furthermore, assume that the set*

$$\{\alpha_1^r\alpha_2^s : 0\le r<R_1,\ 0\le s<S_1\} \tag{8.2}$$

has at least L distinct elements, while the number of distinct elements in the set

$$\{r+s\beta : 0\le r<R_2,\ 0\le s<S_2\} \tag{8.3}$$

is greater than $(K-1)L$. Then at least one of the numbers

$$P(r+s\beta,\alpha_1^r\alpha_2^s), \quad \text{where } 0\le r<R_1+R_2-1,\ 0\le s<S_1+S_2-1, \tag{8.4}$$

is nonzero.

First we demonstrate how Lemma 8.6 with the choice $R_1 = R_2 = S_1 = S_2 = N$ implies Lemma 8.1.

Lemma 8.7. *Under the hypothesis of Theorem* 8.1, *at least one of the two numbers* $\alpha_1 = \alpha$ *and* $\alpha_2 = \alpha^\beta$ *is not a root of unity.*

Proof. If both are, say $\alpha_1 = e^{2\pi i k_1/m}$ and $\alpha_2 = e^{2\pi i k_2/n}$, then

$$\frac{m \ln \alpha}{n\beta \ln \alpha} = \frac{2\pi i k_1}{2\pi i k_2},$$

so that β is rational, which contradicts the hypothesis. □

If α_j is not a root of unity, then the elements α_j^k, $0 \le k < N$, of the set (8.2) are all distinct, so that (8.2) contains at least $N > L = \lfloor N/\ln N \rfloor$ elements. By the irrationality of β all elements in (8.3) (whose number is $N^2 \ge KL > (K-1)L$) are pairwise distinct. Therefore, Lemma 8.6 implies that at least one of the numbers $P(r + s\beta, \alpha_1^r \alpha_2^s)$, $0 \le r, s < 2N - 1$, does not vanish, which contradicts (8.1) and Lemma 8.1 follows. □

It remains to show Lemma 8.6. We will use the following simple observation.

Lemma 8.8. *Let* $k_1, k_2, \dots, k_n$ *be integers,* $0 \le k_1 < k_2 < \cdots < k_n < L$, *and let* $\mathcal{E} \subset \mathbb{C} \setminus \{0\}$ *be a certain (finite) numerical set which contains at least L distinct elements. Then there exist n numbers* $a_1, \dots, a_n \in \mathcal{E}$ *such that the square matrix* $\|a_j^{k_i}\|_{1 \le i,j \le n}$ *is not degenerate.*

First proof. Take L different numbers $b_1, \dots, b_L$ in $\mathcal{E}$ and consider the Vandermonde determinant

$$\det \|b_j^{k-1}\|_{1 \le k,j \le L} = \pm \prod_{1 \le i < j \le L} (b_j - b_i) \ne 0.$$

The rows of the determinant are linearly independent; in particular, the rows with indices $k_1 + 1, k_2 + 1, \dots, k_n + 1$ are linearly independent. Take a nonzero minor spanned by the rows: it corresponds to the required non-degenerate square matrix. □

Second proof. We proceed by induction on n. If $n = 1$, then a_1 can be taken any in $\mathcal{E} \subset \mathbb{C} \setminus \{0\}$.

Assume that we have already managed a collection $a_1, \ldots, a_{n-1} \in \mathcal{E}$ such that the determinant $A = \det \|a_j^{k_i}\|_{1 \le i,j \le n-1}$ does not vanish. Consider the polynomial

$$P(z) = \det \begin{Vmatrix} a_1^{k_1} & \ldots & a_{n-1}^{k_1} & z^{k_1} \\ \ldots & \ldots & \ldots & \ldots \\ a_1^{k_n} & \ldots & a_{n-1}^{k_n} & z^{k_n} \end{Vmatrix}. \tag{8.5}$$

The substitutions $z = a_j$, $j = 1, \ldots, n-1$, obviously make it zero, so that

$$P(z) = (z - a_1) \cdots (z - a_{n-1}) Q(z)$$

for a certain polynomial $Q(z)$. Expanding the determinant (8.5) along the last column we see that $P(z) = Az^{k_n} +$ lower degree terms; so that $\deg P = k_n < L$ and $\deg Q = \deg P - (n-1) < L - n + 1$. The set $\mathcal{E} \setminus \{a_1, \ldots, a_{n-1}\}$ contains at least $L - n + 1 > \deg Q$ elements, so that for at least one of them, say a_n, we have $Q(a_n) \ne 0$. This proves the induction step and completes the proof of the lemma. □

Proof of Lemma 8.6. We proceed the proof by contradiction, assuming that all the numbers in (8.4) are zero.

Expand the polynomial $P(x, y)$ in powers of y, writing only those monomials y^k whose coefficients are nonzero:

$$P(x, y) = \sum_{i=1}^{n} Q_i(x) y^{k_i}, \quad Q_i(x) \not\equiv 0 \text{ for } i = 1, \ldots, n,$$
$$0 \le k_1 < k_2 < \cdots < k_n < L.$$

Define the set

$$\mathcal{E} = \{\alpha_1^r \alpha_2^s : 0 \le r < R_1,\ 0 \le s < S_1\} \subset \mathbb{C} \setminus \{0\};$$

the number of distinct elements in $\mathcal{E}$ is at least L by the hypothesis. In accordance with Lemma 8.8 choose an n-element subset $\mathcal{L} = \{(r, s)\} \subset \mathcal{E}$ such that

$$B = \det \|(\alpha_1^r \alpha_2^s)^{k_i}\|_{i=1,\ldots,n;\ (r,s) \in \mathcal{L}} \ne 0.$$

Consider the polynomials

$$\begin{aligned} P_{r,s}(x, y) &= P(x + r + s\beta, \alpha_1^r \alpha_2^s y) \\ &= \sum_{i=1}^{n} Q_i(x + r + s\beta)(\alpha_1^r \alpha_2^s)^{k_i} y^{k_i}, \quad (r, s) \in \mathcal{L}. \end{aligned}$$

By our assumption

$$P_{r,s}(r' + s'\beta, \alpha_1^{r'}\alpha_2^{s'}) = P((r + r') + (s + s')\beta, \alpha_1^{r+r'}\alpha_2^{s+s'}) = 0 \\ \text{for all } 0 \le r' < R_2,\ 0 \le s' < S_2. \tag{8.6}$$

Finally, define

$$\Delta(x) = \det \|Q_i(x + r + s\beta)(\alpha_1^r\alpha_2^s)^{k_i}\|_{i=1,\dots,n;\ (r,s)\in\mathcal{L}}.$$

Each of the polynomials $Q_i(x)$ is given in the form $Q_i(x) = b_i x^{m_i} +$ lower degree terms, where b_i are nonzero. Therefore, expanding the determinant $\Delta(x)$ we obtain $\Delta(x) = Ax^{m_1+\cdots+m_n} +$ lower degree terms, where

$$A = \det \|b_i(\alpha_1^r\alpha_2^s)^{k_i}\|_{i=1,\dots,n;\ (r,s)\in\mathcal{L}} = b_1 \cdots b_n \cdot B \ne 0.$$

Thus, $\Delta(x) \not\equiv 0$ and $\deg \Delta(x) = m_1 + \cdots + m_n$.

Consider now the system of n linear equations

$$\sum_{i=1}^{n} Q_i(x + r + s\beta)(\alpha_1^r\alpha_2^s)^{k_i} y^{k_i} = P_{r,s}(x, y), \quad (r, s) \in \mathcal{L},$$

in which y^{k_i} are counted as n unknowns. The determinant of the system is exactly $\Delta(x)$. Solving the system by Cramer's rule we get

$$\Delta(x) \cdot y^{k_i} = \Delta_i \quad \text{for } i = 1, \dots, n, \tag{8.7}$$

where the determinant Δ_i is obtained from Δ by replacing the ith column with $\|P_{r,s}(x, y)\|_{(r,s)\in\mathcal{L}}$. Substituting $x = r' + s'\beta$ and $y = \alpha_1^{r'}\alpha_2^{s'}$, where $r' = 0, 1, \dots, R_2 - 1$ and $s' = 0, 1, \dots, S_2$, makes the latter column vanishing, so that the minors Δ_i in (8.7) vanish as well:

$$\Delta(r' + s'\beta) \cdot (\alpha_1^{r'}\alpha_2^{s'})^{k_i} = 0, \quad i = 1, \dots, n, \\ \text{for all } 0 \le r' < R_2,\ 0 \le s' < S_2.$$

This, in turn, implies that $\Delta(r' + s'\beta) = 0$ for all $0 \le r' < R_2$ and $0 \le s' < S_2$. By the hypothesis, there are more than $(K - 1)L$ distinct numbers in the set (8.3), so that the number of zeros of the polynomial $\Delta(x)$ must be greater than $(K - 1)L$, hence $\deg \Delta(x) > (K - 1)L$. On the other hand,

$$\deg \Delta(x) = m_1 + \cdots + m_n \le (K - 1)n < (K - 1)L.$$

The contradiction we arrived at, shows that at least one of the numbers in (8.4) does not vanish. □

8.4 Schneider's proof of the principal theorem

Assume on contrary that the numbers α, β and $\gamma = \alpha^\beta$ are algebraic. Since this assumption is equivalent to the algebraicity of $\alpha_1 = \alpha^\beta$, $\beta_1 = 1/\beta$ and $\gamma_1 = \alpha_1^{\beta_1} = \alpha$, we can swap the original set with this one and proceed with the proof for the newer set instead. We will choose one of the two sets, α, β, γ and $\alpha_1, \beta_1, \gamma_1$, and call it α, β, γ in what follows, a set for which α is not a root of unity (such a choice is guaranteed by Lemma 8.7).

As in the previous proof we choose the parameters $K = \lfloor N \ln N \rfloor$ and $L = \lfloor N/\ln N \rfloor$. The field generated by algebraic numbers α, β, γ is denoted by F, while its ring of integers is denoted by $\mathbb{Z}_F$. Notation $\overline{|a|}$ (the house) from Section 6.3 stands for the maximum of the absolute values of $a \in F$ and all its conjugates in F.

Lemma 8.9. *For any sufficiently large N, there exists a function*

$$f(z) = \sum_{l=0}^{L-1} P_l(z)\alpha^{lz} \not\equiv 0, \quad P_l(z) = \sum_{k=0}^{K-1} A_{lk} z^k, \quad l = 0, 1, \ldots, L-1, \tag{8.8}$$

with coefficients $A_{lk} \in \mathbb{Z}_F$ (not all simultaneously zero) such that

$$\overline{|A_{lk}|} \le e^{N^2/\sqrt{\ln N}} \tag{8.9}$$

and

$$f(r + s\beta) = 0 \quad \text{for all } 0 \le r, s < M = \left\lfloor \frac{1}{2} N \right\rfloor. \tag{8.10}$$

Remark. Conditions (8.10) corresponds to the linear system

$$\sum_{k=0}^{K-1} \sum_{l=0}^{L-1} A_{lk} \cdot (r + s\beta)^k \alpha^{l(r+s\beta)} = 0, \quad 0 \le r, s < M, \tag{8.11}$$

with A_{lk} as unknowns, $0 \le l < L-1$, $0 \le k < K$. Since $\alpha^{l(r+s\beta)} = (\alpha_1^r \alpha_2^s)^l$, the matrix of the linear system is exactly the matrix $\mathcal{M}$ from Section 8.1 when $M = 2N$. In other words, our proof of Theorem 8.1 with interpolation determinants used the fact that a nonzero minor of maximal order KL of the matrix corresponding to the system (8.11) is itself sufficiently small, and at the same time it is a polynomial in the numbers α, β, γ under consideration. It is this circumstance which underlies many other proofs by Laurent's method of interpolation determinants: instead of solving a linear system, one investigates a nonzero minor of the corresponding matrix.

Proof. To solve the system (8.11) we use the following result.

Exercise 8.2 (Siegel lemma). Show that the system of linear equations

$$\sum_{j=1}^{q} a_{ij}x_j = 0, \quad i = 1, \ldots, p,$$

with coefficients $a_{ij} \in \mathbb{Z}_F$, $\lceil a_{ij} \rceil \leq A$, in $q > p$ unknowns $x_1, \ldots, x_q$, possesses a non-trivial solution

$$x_j \in \mathbb{Z}_F, \ \lceil x_j \rceil \leq c(cqA)^{p/(q-p)}, \quad j = 1, \ldots, q,$$

where the positive constant c depends only on the field F.

Hint. Prove the statement first for the case $F = \mathbb{Q}$ using the pigeon hole principle (as in the proof of Theorem 6.3); then reduce the general case to this particular situation. □

In our case the number of unknowns, $q = KL \sim N^2$, is greater than the number of equations, $p = M^2 \sim N^2/4$. If positive integer d is such that $d\alpha, d\beta, d\gamma \in \mathbb{Z}_F$, then after multiplying all equations in (8.11) by d^{K+2LM} we obtain a system of linear equations with coefficients from $\mathbb{Z}_F$. The coefficients are then estimated as follows:

$$\begin{aligned} \lceil d^{K+2LM}(r+s\beta)^k\alpha^{lr}\gamma^{ls} \rceil &\leq d^{K+2LM}(2M)^K \lceil\beta\rceil^K \lceil\alpha\rceil^{LM} \lceil\gamma\rceil^{LM} \\ &\leq e^{c_1 N^2/\ln N} = A. \end{aligned}$$

In addition, $p/(q-p) \sim 1/3$ as $N \to \infty$. By the Siegel lemma there exists a non-trivial solution of (8.11) in unknowns $A_{lk} \in \mathbb{Z}_F$ such that

$$\lceil A_{lk} \rceil \leq c(cN^2 e^{c_1 N^2/\ln N})^{1/3} = e^{c_2 N^2/\ln N} < e^{N^2/\sqrt{\ln N}}$$

for any sufficiently large N. This proves Lemma 8.9. □

From Lemma 8.6 established above, with the same choice $R_1 = R_2 = S_1 = S_2 = N$, it follows that there exists a pair r_0, s_0, $0 \leq r_0, s_0 < 2N$, such that $f(r_0 + s_0\beta) \neq 0$ for the function $f(z)$ constructed in Lemma 8.9. Denote $\delta = f(r_0 + s_0\beta) \neq 0$; we will estimate the number from above and from below.

An estimate of δ from above. Consider the function

$$g(z) = \frac{f(z)}{\prod_{0 \leq r,s < M}(z - r - s\beta)}.$$

By (8.10) it is analytic on $\mathbb{C}$. We have $|r_0 + s_0\beta| \leq 2N(1 + |\beta|)$. Define the radii $\rho = 2N(1 + |\beta|)$ and $R = 5\rho$, and apply the maximum modulus principle:

$$
\begin{aligned}
|\delta| = |f(r_0 + s_0\beta)| &\le |g|_\rho \cdot \left| \prod_{0\le r,s<M} (z - r - s\beta) \right|_\rho \\
&\le |g|_R \cdot \left| \prod_{0\le r,s<M} (z - r - s\beta) \right|_\rho \\
&\le |f|_R \cdot \prod_{0\le r,s<M} \left| \frac{z_1 - r - s\beta}{z_2 - r - s\beta} \right|,
\end{aligned}
\tag{8.12}
$$

where the points z_1, $|z_1| = \rho$, and z_2, $|z_2| = R$, are taken in such a way that the maximum moduli of $\prod_{0\le r,s<M}(z - r - s\beta)$ on the circles $|z| = \rho$ and $|z| = R$ are attained at them. Then

$$
\left.
\begin{aligned}
|z_1 - r - s\beta| &\le |z_1| + |r + s\beta| \le 2\rho, \\
|z_2 - r - s\beta| &\ge |z_2| - |r + s\beta| \ge R - \rho,
\end{aligned}
\right.
\qquad 0 \le r, s < M;
$$

therefore, the estimate in (8.12) can be continued as follows:

$$
|\delta| \le |f|_R \cdot \left(\frac{2\rho}{R - \rho} \right)^{M^2} = |f|_R \cdot 2^{-M^2}.
$$

Now, using definition (8.8) of the function $f(z)$ and the estimates (8.9), we get

$$
|f|_R \le KL \cdot e^{N^2/\sqrt{\ln N}} \cdot R^K \cdot e^{LR|\ln\alpha|} \le e^{2N^2/\sqrt{\ln N}}
$$

implying

$$
|\delta| < e^{2N^2/\sqrt{\ln N}} \cdot e^{-(N^2\ln 2)/4} < e^{-N^2/10} \tag{8.13}
$$

for all sufficiently large N.

An estimate of δ from below. Write δ as $\delta = Q(\alpha, \beta, \gamma)$, where

$$
Q(x, y, z) = \sum_k \sum_l A_{lk}(r_0 + s_0 y)^k x^{lr_0} z^{ls_0} \in \mathbb{Z}_F[x, y, z].
$$

By Liouville's theorem the estimate

$$
\ln|Q(\alpha, \beta, \gamma)| \ge -c(\deg Q + \ln\overline{|Q|})
$$

holds, where the constant $c > 0$ depends only on the field F and $\overline{|Q|}$ denotes the sum of the houses of the coefficients of Q. In our situation

$$
\deg Q \le 3LM \le \frac{3N^2}{\ln N},
$$

$$
\ln\overline{|Q|} \le \frac{N^2}{\sqrt{\ln N}} + K\ln\bigl(M(1 + \overline{|\beta|})\bigr) \le \frac{2N^2}{\sqrt{\ln N}},
$$

so that

$$
|\delta| = |Q(\alpha, \beta, \gamma)| \ge e^{-3cN^2/\sqrt{\ln N}} \tag{8.14}
$$

for all sufficiently large N.

The estimates (8.13) and (8.14) contradict each other. Hence our assumption about the simultaneous algebraicity of α, β and $\gamma = \alpha^\beta$ is false. This completes the proof of Theorem 8.1.

Chapter notes

The resolutions of Hilbert's seventh problem were published independently (and almost simultaneously) in 1934. In spite of the similarity of Gelfond's and Schneider's methods, constructions of the auxiliary function were quite different (Lemma 8.9 highlights Schneider's choice), and this difference played a crucial role in the later development of the theory of transcendental numbers. For example, Schneider's version was used by Schneider himself [70] to prove results about the transcendence of values of elliptic functions and elliptic modular functions; in a general form the results are known as the Schneider–Lang theorem [53]. The development of Gelfond's method culminated in what is called Baker's theorem — effective lower bounds for the absolute value of linear combinations of logarithms of algebraic numbers [7, 19, 78].

Laurent's method of interpolation determinants is considerably young [54,55] but demonstrates a significant power in applications to transcendental numbers. One of its outcomes is sharp bounds for linear forms in two logarithms (of algebraic numbers) [56], which is of particular importance for applications to diophantine equations.

Chapter 9

Schinzel–Zassenhaus conjecture

The aim of this chapter is to outline a remarkable proof of the Schinzel–Zassenhaus conjecture given at the end of 2019 by V. Dimitrov in [26].

Theorem 9.1 (Dimitrov [26]). *For an algebraic integer α of degree d, not a root of unity, its house — the maximum modulus of its conjugates (including α itself) — satisfies* $\overline{\lceil\alpha\rceil} \ge 2^{1/(4d)}$.

This indeed answers the 1965 suspicion of Schinzel and Zassenhaus in [69] about the bound $\overline{\lceil\alpha\rceil} \ge 1 + c/d$ for some absolute constant $c > 0$. Theorem 9.1 allows one to take $c = (\log 2)/4$. The earlier recorded partial resolutions of the Schinzel–Zassenhaus conjecture all appealed to related resolutions of Lehmer's problem [18].

Dimitrov's proof is based on the following ingredients given in Propositions 9.1, 9.2, 9.3 and 9.4 below.

Proposition 9.1 (Dimitrov [26]). *For an algebraic integer α, denote by $P(x) = \prod_{j=1}^{d}(x - \alpha_j) \in \mathbb{Z}[x]$ its minimal (monic!) polynomial. Introduce additionally the polynomials*

$$P_2(x) = \prod_{j=1}^{d}(x - \alpha_j^2) \in \mathbb{Z}[x] \quad \text{and} \quad P_4(x) = \prod_{j=1}^{d}(x - \alpha_j^4) \in \mathbb{Z}[x],$$

and assume that $P_2(x)$ is irreducible over $\mathbb{Z}$. Then

$$f(z) = \sqrt{P_2(z)P_4(z)/z^{2d}} \in 1 + z^{-1}\mathbb{Z}[[z^{-1}]],$$

and $f(z)$ is rational if and only if $P(x)$ is cyclotomic.

Notice that the statement translates cyclotomicity of $P(x)$ into a rationality criterion for $f(z)$; it sieves out those α that are roots of unity.

Proposition 9.2 (Pólya [62]). *For a compact K in $\mathbb{C}$, let $\hat{K}$ be a simply connected compact containing K. Assume that the function $f(z)$ is analytic on $\overline{\mathbb{C}}\setminus\hat{K}$ (that is, on a connected component of the complement of K which contains ∞) and possesses the expansion*

$$f(z)=\sum_{k=0}^{\infty}\frac{a_k}{z^k}$$

at ∞. Define the Hankel determinants $A_n=\det_{0\le j,\ell<n}(a_{j+\ell})$ for $n=1,2,\dots$. Then

$$\limsup_{n\to\infty}|A_n|^{1/n^2}\le t(\hat{K}),$$

the transfinite diameter of $\hat{K}$.

The transfinite diameter $t(K)$ of a compact $K\subset\mathbb{C}$, also known as the (logarithmic) capacity of K or the Chebyshev–Fekete constant of K, is defined in Section 9.2.

The next statement is known as Kronecker's rationality criterion.

Proposition 9.3 (Kronecker [50, pp. 566–567]). *Let $f(x)=\sum_{n=0}^{\infty}a_nx^n\in\mathbb{C}[[x]]$ be a formal power series. Then $f(x)$ is a quotient of two polynomials (in other words, represents a rational function) if and only if $A_n=0$ for all $n\ge n_1$, where $A_n=\det_{0\le j,\ell<n}(a_{j+\ell})$.*

The following result is a consequence of Dubinin's solution of a problem of Gonchar. In order to state it, define a *hedgehog* with vertices $\beta_1,\dots,\beta_d\in\mathbb{C}^\times$, notation $K(\beta_1,\dots,\beta_d)\subset\mathbb{C}$, to be the union of the d closed radial segments $[0,\beta_j]$ joining the origin 0 to the points β_j in the complex plane, for $j=1,\dots,d$. Note that a hedgehog K is already simply connected, so that Proposition 9.2 applies to $\hat{K}=K$.

Proposition 9.4 (Dubinin [27]). *The hedgehog $K=K(\beta_1,\dots,\beta_d)\subset\mathbb{C}$ has transfinite diameter $t(K)$ at most*

$$\left(\frac{1}{4}\max_{1\le j\le d}|\beta_j|^d\right)^{1/d}=4^{-1/d}\max_{1\le j\le d}|\beta_j|.$$

Proof of Theorem 9.1. Throughout the proof we assume that α is not a root of unity, so that $\overline{|\alpha|}>1$.

We proceed by induction on degree d; the estimate $\overline{|\alpha|}\ge 2>2^{1/4}$ is clearly true when $d=1$, and we assume that the theorem is shown for all algebraic integers of degree less than given $d>1$.

If $P_2(x)$ is reducible, then α^2 has degree $d/2$, so that it satisfies $\overline{|\alpha^2|} \geq 2^{1/(2d)}$ by the induction hypothesis. Since $\overline{|\alpha^2|} = \overline{|\alpha|}^2$, we get the desired inequality. Therefore, we can assume that $P_2(x)$ is irreducible, hence by Proposition 9.1 the function

$$f(z) = \sqrt{P_2(z)P_4(z)/z^{2d}} = 1 + \sum_{k=1}^{\infty} \frac{a_k}{z^k}$$

has all coefficients at ∞ integral and is irrational. Then by Proposition 9.3 infinitely many of the Hankel determinants $A_n = \det_{0 \leq i,j < n}(a_{i+j}) \in \mathbb{Z}$ do not vanish; in particular, all those satisfy $|A_n| \geq 1$. By Proposition 9.2 this means that $t(K) \geq 1$, where K is the hedgehog spanned by α^2, α^4 and all their conjugates; in particular, Proposition 9.4 implies that $t(K) \leq 4^{-1/(2d)}\overline{|\alpha|}^4$. Combining the two estimates implies $\overline{|\alpha|}^4 \geq 4^{1/(2d)} = 2^{1/d}$ and leads to the inequality claimed. □

In the remaining part we prove Propositions 9.1–9.3 and give some intuition behind Proposition 9.4; each section takes care of the corresponding proposition.

9.1 Dimitrov's cyclotomicity criterion

Notably, Fermat's little theorem $a^p \equiv a \pmod{p}$ for all $a \in \mathbb{Z}$ and primes p generalises to Euler's congruence

$$a^{p^r} \equiv a^{p^{r-1}} \pmod{p^r}, \quad \text{where } r = 1, 2, \ldots,$$

and further to the Gauss congruence

$$\sum_{d|m} \mu\left(\frac{m}{d}\right) a^d \equiv 0 \pmod{m}, \tag{9.1}$$

where $\mu(\,\cdot\,)$ is the Möbius function, valid for all positive integers m; see [72, 83]. The validity for $m = p^r = 2^2$ can be performed by hand: if a is even then both a^4 and a^2 are divisible by 4; if $a = 2k + 1$ then

$$a^4 - a^2 = (2k+1)^4 - (2k+1)^2 = 16k^4 + 32k^3 + 20k^2 + 4k \equiv 0 \pmod{4}.$$

Exercise 9.1. Given $a \in \mathbb{Z}$, prove the Gauss congruence (9.1) for any $m = 1, 2, \ldots$.

Exercise 9.2. Let $\{a_m\}_{m \geq 1}$ be a sequence of integers. Then the following two conditions are equivalent:

(i) for every $m = 1, 2, \dots$,

$$\sum_{d|m} \mu\left(\frac{m}{d}\right) a_d \equiv 0 \pmod{m},$$

that is, the sequence satisfies Gauss congruences; and

(ii) for all $n, s \ge 1$ and all primes p, we have $a_{p^s n} \equiv a_{p^{s-1} n} \pmod{p^s}$.

Lemma 9.1. *For a monic polynomial $P(x) \in \mathbb{Z}[x]$, we have the congruence*

$$P_4(x) \equiv P_2(x) \pmod{4},$$

where the congruence is understood as the congruence of the corresponding individual coefficients of polynomials.

Proof. Every symmetric function on the zeros $\alpha_1, \dots, \alpha_d$ of $P(x) = x^d - e_1 x^{d-1} + e_2 x^{d-2} + \cdots + (-1)^d e_d$ can be written as a polynomial in the symmetric functions $e_1, e_2, \dots, e_d$. Such representations for the sums of powers of the zeros, $s_k = \sum_{j=1}^d \alpha_j^k$, are known as the Newton–Girard identities; explicitly, we have (easily derivable!)

$$\begin{gathered} s_1 = e_1, \quad s_2 = e_1^2 - 2e_2, \quad s_3 = e_1^3 - 3e_1e_2 + 3e_3, \\ s_4 = e_1^4 - 4e_1^2e_2 + 4e_1e_3 + 2e_2^2 - 4e_4, \end{gathered} \tag{9.2}$$

and so on. The coefficients of these polynomials are always integral. Now, since $e_1, e_2, e_3, e_4 \in \mathbb{Z}$ and $e_1^4 \equiv e_1^2 \pmod{4}$ (by the above), $e_2^2 \equiv -e_2 \pmod{2}$, we deduce from (9.2) (from the expressions for s_2 and s_4 only) that $s_4 \equiv s_2 \pmod{4}$, that is,

$$e_1(\alpha_1^4, \dots, \alpha_d^4) \equiv e_1(\alpha_1^2, \dots, \alpha_d^2) \pmod{4}.$$

By replacing the original system of zeros with $\{\alpha_{j_1} \cdots \alpha_{j_k} : 1 \le j_1 < \cdots < j_k \le d\}$, hence the original polynomial with the corresponding one (of possibly higher degree!), still monic and with integral coefficients, and applying the same argument we deduce that

$$e_k(\alpha_1^4, \dots, \alpha_d^4) \equiv e_k(\alpha_1^2, \dots, \alpha_d^2) \pmod{4}$$

for $k = 2, 3, \dots, d$ as well. □

Proof of Proposition 9.1. Observe that

$$\begin{aligned} \sqrt{1+4X} &= \sum_{n=0}^{\infty} \frac{\prod_{j=0}^{n-1}(\frac{1}{2} - j)}{n!}(4X)^n \\ &= 1 + 2\sum_{n=1}^{\infty}(-1)^{n-1}C_{n-1}X^n \in 1 + X\mathbb{Z}[[X]], \end{aligned} \tag{9.3}$$

where the integrality of the Catalan numbers $C_n = \binom{2n}{n}/(n+1)$ has been discussed in Exercise 1.4. From Lemma 9.1 we have $P_2(x)P_4(x) = P_2(x)(P_2(x)+4Q(x))$ for some polynomial $Q(x) \in \mathbb{Z}[x]$ of degree less than d. Thus,

$$\sqrt{P_2(x)P_4(x)} = P_2(x)\sqrt{1 + 4Q(x)/P_2(x)},$$

and the result follows from application of (9.3) with $X = Q(z)/P_2(z) \in 1 + z^{-1}\mathbb{Z}[[z^{-1}]]$.

The rationality of $f(z)$ would mean that

$$P_2(x)P_4(x) = \prod_{j=1}^{d}(x - \alpha_j^2)(x - \alpha_j^4)$$

is a square in $\mathbb{Z}[x]$. Since $P_2(x)$ is irreducible by the hypothesis, the numbers $\alpha_1^2, \dots, \alpha_d^2$ are pairwise distinct, hence each of them pairs up with some zero of $P_4(x)$: $\alpha_j^2 = \alpha_{\sigma(j)}^4$ for each $j = 1, \dots, d$. The mapping σ is clearly a permutation of the indices of $\alpha_1, \dots, \alpha_d$. Iterating the identity,

$$\alpha_j^2 = (\alpha_{\sigma(j)}^2)^2 = \alpha_{\sigma^2(j)}^8 = \cdots = \alpha_{\sigma^k(j)}^{2^{k+1}} \quad \text{for } k = 1, 2, \dots,$$

and using the fact that σ^k is the identity for some k, we conclude that $\alpha_j^2 = \alpha_j^{2^{k+1}}$ implying that each α_j is a zero of the polynomial $x^{2^{k+1}-2} - 1$. In particular, $\alpha_1, \dots, \alpha_d$ are roots of unity, thus our polynomial $P(x)$ is cyclotomic. □

9.2 Hankel determinants and transfinite diameter

Recall that the Vandermonde determinant evaluation

$$V(z_1, \dots, z_n) = \det_{1 \le j, \ell \le n}(z_j^{\ell-1}) = \prod_{1 \le j < \ell \le n}(z_\ell - z_j).$$

Lemma 9.2 (Fekete [29]). *Let K be a compact in $\mathbb{C}$ containing infinitely many points. Denote by M_n the maximum of the quantity $|V(z_1, \dots, z_n)|$ as $z_1, \dots, z_n$ run through the set K. Then the limit*

$$t(K) = \lim_{n \to \infty} M_n^{2/(n(n-1))}$$

exists.

The limit $t(K)$ is called the *transfinite diameter* of K. Observe that the definition implies that $t(K) \le t(K')$ whenever we have $K \subset K'$ for two compacts K and K' in $\mathbb{C}$.

Proof. We borrow the argument from [48, Problem 1.10]. Given $n \ge 1$, assume that the maximum M_n of $|V(z_1, \dots, z_n)|$ is attained at $\zeta_1, \dots, \zeta_n \in K$. Because

$$\frac{V(\zeta_1, \dots, \zeta_n)}{V(\zeta_1, \dots, \zeta_{n-1})} = (\zeta_n - \zeta_1) \cdots (\zeta_n - \zeta_{n-1}),$$

we conclude that

$$\frac{M_n}{M_{n-1}} \le |\zeta_n - \zeta_1| \cdots |\zeta_n - \zeta_{n-1}|.$$

With the same argument used for ζ_n replaced with any other ζ_j we find out that

$$\left(\frac{M_n}{M_{n-1}}\right)^n \le \prod_{\substack{j,\ell=1 \\ j \ne \ell}}^{n} |\zeta_\ell - \zeta_j| = M_n^2,$$

equivalently, $M_n^{1/n} \le M_{n-1}^{1/(n-2)}$ implying that $M_n^{2/(n(n-1))}$ is monotone decreasing. □

Proof of Proposition 9.2. Choose γ to be a piecewise smooth closed contour in $\overline{\mathbb{C}} \setminus \hat{K}$, which is positively oriented with respect to ∞. The function $f(z)$ is analytic within the exterior of γ, so that Cauchy integral formula applies and we obtain

$$a_k = \frac{1}{2\pi i} \int_\gamma f(z) z^{k-1} \, \mathrm{d}z$$

for the coefficients of the expansion of $f(z)$ at infinity. Therefore,

$$\begin{aligned} A_n = \det_{1 \le j,\ell \le n} (a_{j+\ell-2}) &= \det_{1 \le j,\ell \le n} \left(\int_\gamma f(z_j) z_j^{j+\ell-2} \, \mathrm{d}z_j \right) \\ &= \int \cdots \int_{\gamma^n} \det_{1 \le j,\ell \le n} \left(z_j^{j+\ell-2} \right) \cdot \prod_{j=1}^n f(z_j) \, \mathrm{d}z_j \\ &= \int \cdots \int_{\gamma^n} \prod_{j=1}^n z_j^{j-1} \cdot \det_{1 \le j,\ell \le n} \left(z_j^{\ell-1} \right) \cdot \prod_{j=1}^n f(z_j) \, \mathrm{d}z_j \\ &= \int \cdots \int_{\gamma^n} \prod_{j=1}^n z_j^{j-1} \cdot V(z_1, \dots, z_n) \cdot \prod_{j=1}^n f(z_j) \, \mathrm{d}z_j. \end{aligned}$$

Now using

$$\sum_{\sigma \in \mathfrak{S}_n} \operatorname{sgn}(\sigma) \prod_{j=1}^n z_{\sigma(j)}^{j-1} = \det_{1 \le j,\ell \le n} \left(z_j^{\ell-1} \right) = V(z_1, \dots, z_n)$$

and

$$V(z_{\sigma(1)}, \dots, z_{\sigma(n)}) = \operatorname{sgn}(\sigma)\, V(z_1, \dots, z_n) \quad \text{for } \sigma \in \mathfrak{S}_n,$$

and averaging the resulting n-tuple integral for A_n over all substitutions σ of the symmetric group $\mathfrak{S}_n$ we deduce that

$$A_n = \frac{1}{n!} \int \cdots \int_{\gamma^n} V(z_1, \dots, z_n)^2 \cdot \prod_{j=1}^{n} f(z_j)\, \mathrm{d}z_j,$$

so that

$$|A_n| \le \frac{(M(\gamma)L(\gamma))^n}{n!} \cdot \max_{z_1,\dots,z_n \in \gamma} |V(z_1, \dots, z_n)|^2$$

where $M(\gamma)$ is the maximum of $|f(z)|$ on γ and $L(\gamma)$ is the length of γ. Since this is valid for *any* contour γ enclosing the compact $\hat{K}$, we can record the resulting inequality in the form

$$|A_n| \le \frac{C^n}{n!} \cdot \max_{z_1,\dots,z_n \in \hat{K}} |V(z_1, \dots, z_n)|^2$$

for some positive constant C independent of n. Raising both sides to the power $1/n^2$, taking the limit superior as $n \to \infty$ and applying Lemma 9.2, the statement of Proposition 9.2 follows. □

9.3 Kronecker's rationality criterion

Proof of Proposition 9.3. First notice that a power series $f(x)$ represents a rational function if and only the sequence of its coefficients satisfies a recurrence relation

$$c_0 a_n + c_1 a_{n+1} + \cdots + c_m a_{n+m} = 0 \quad \text{for all } n \ge n_0$$

with constant coefficients $c_0, c_1, \dots, c_m$. If such a relation is available and $n > n_0 + m$ is arbitrary, then the columns starting with $a_{n_0}, a_{n_0+1}, \dots, a_{n_0+m}$ in the determinant A_n are linearly dependent, hence $A_n = 0$ for all such n.

We are left to show that $A_n = 0$ for all $n \ge n_1$ implies a recurrence relation for a_n with constant coefficients. Choose m to be such that $A_m \ne 0$ while $A_n = 0$ for all $n > m$. The former condition implies that the first m columns of the matrix for A_{m+1} are linearly independent. On the other hand, $A_{m+1} = 0$ means that the last column of the determinant A_{m+1} is a linear combination of all previous ones:

$$c_0 a_n + c_1 a_{n+1} + \cdots + c_{m-1} a_{n+m-1} + a_{n+m} = 0 \quad \text{for } n = 0, 1, \dots, m.$$

We will show that the equality holds true for *all* n, in other words, that $b_n = 0$ for all $n \ge 0$ where

$$b_n = c_0 a_n + c_1 a_{n+1} + \cdots + c_{m-1} a_{n+m-1} + a_{n+m}.$$

For $n > m$, assume that $b_0 = b_1 = \cdots = b_{n-1} = 0$ is already shown, write

$$A_{n+1} = \det \left(\begin{array}{ccc|ccc} a_0 & \cdots & a_{m-1} & a_m & \cdots & a_n \\ \vdots & \ddots & \vdots & \vdots & & \vdots \\ a_{m-1} & \cdots & a_{2m-2} & \vdots & & \vdots \\ \hline a_m & \cdots & \cdots & & & a_{n+m} \\ \vdots & & & & \cdot^{\cdot^{\cdot}} & \vdots \\ a_n & \cdots & \cdots & a_{n+m} & \cdots & a_{2n} \end{array} \right)$$

and add to each column in the right part the linear combination of m preceding columns with the corresponding coefficients $c_0, c_1, \dots, c_{m-1}$. Then

$$\begin{aligned} A_{n+1} &= \det \left(\begin{array}{ccc|ccc} a_0 & \cdots & a_{m-1} & b_0 & \cdots & b_{n-m} \\ \vdots & \ddots & \vdots & \vdots & & \vdots \\ a_{m-1} & \cdots & a_{2m-2} & \vdots & & \vdots \\ \hline a_m & \cdots & \cdots & & & b_n \\ \vdots & & & & \cdot^{\cdot^{\cdot}} & \vdots \\ a_n & \cdots & \cdots & b_n & \cdots & b_{2n-m} \end{array} \right) \\ &= (-1)^{n-m} A_m \cdot b_n^{n-m+1}, \end{aligned}$$

because all the entries above the anti-subdiagonal $b_n \ \dots \ b_n$ vanish. Using now $A_{n+1} = 0$ and $A_m \ne 0$, we conclude that $b_n = 0$ as required. □

The next result is a variation of the rationality criterion from Proposition 9.3 also established by Kronecker.

Exercise 9.3. With a formal power series $f(x) = \sum_{n=0}^{\infty} a_n x^n \in \mathbb{C}[[x]]$ associate general $m \times m$ Hankel determinants

$$H_{n,m} = \det_{0 \le j,\ell \le m} (a_{n+j+\ell}).$$

(a) Show that $f(x)$ is a rational function if and only if $H_{n,m} = 0$ for *some* m and *all* $n \ge n_1$.

(b) For $n, m = 1, 2, \dots$, prove the identity

$$H_{n-1,m} H_{n+1,m} - H_{n-1,m+1} H_{n+1,m-1} = H_{n,m}^2.$$

9.4 Transfinite diameter of a hedgehog

For a compact $K \subset \mathbb{C}$, its nth Chebyshev polynomial $T_n(z)$ is a degree n monic polynomial which minimises the sup-norm

$$\|f\|_K = \sup_{z \in K} |f(z)|$$

over all degree n monic polynomials. A simple argument (see [24, p. 208]) shows that the Chebyshev polynomial $T_n(z)$ is unique. As in Lemma 9.2, denote M_n to be the maximum of $|V(z_1, \dots, z_n)|$ over all $z_1, \dots, z_n$ in K and assume that it is attained at $\zeta_1, \dots, \zeta_n \in K$. Since

$$\begin{aligned} V(\zeta_1, \dots, \zeta_n) &= \det \begin{pmatrix} 1 & \zeta_1 & \cdots & \zeta_1^{n-2} & \zeta_1^{n-1} \\ 1 & \zeta_2 & \cdots & \zeta_2^{n-2} & \zeta_2^{n-1} \\ \vdots & \vdots & \ddots & \vdots & \vdots \\ 1 & \zeta_n & \cdots & \zeta_n^{n-2} & \zeta_n^{n-1} \end{pmatrix} \\ &= \det \begin{pmatrix} 1 & \zeta_1 & \cdots & \zeta_1^{n-2} & T_{n-1}(\zeta_1) \\ 1 & \zeta_2 & \cdots & \zeta_2^{n-2} & T_{n-1}(\zeta_2) \\ \vdots & \vdots & \ddots & \vdots & \vdots \\ 1 & \zeta_n & \cdots & \zeta_n^{n-2} & T_{n-1}(\zeta_n) \end{pmatrix}, \end{aligned}$$

expanding along the last column leads to

$$\begin{aligned} M_n &\le |T_{n-1}(\zeta_1)| \cdot |V(\zeta_2, \dots, \zeta_n)| + \cdots + |T_{n-1}(\zeta_n)| \cdot |V(\zeta_1, \dots, \zeta_{n-1})| \\ &\le n\|T_{n-1}\|_K \cdot M_{n-1}. \end{aligned}$$

This means that

$$M_n \le n! \cdot \prod_{j=1}^{n-1} \|T_j\|_K,$$

so that if $\limsup_{n\to\infty} \|T_n\|^{1/n} = t^*(K)$ (essentially the Chebyshev constant of the compact K), then

$$M_n \le n! \cdot C^{n-1} t^*(K)^{1+2+\cdots+(n-1)} = n! \cdot C^{n-1} t^*(K)^{n(n-1)/2}$$

for some $C = C(K)$ independent of n, implying $t(K) \le t^*(K)$.

Furthermore, for the interval $K = [a, b] \subset \mathbb{R}$, it is known that

$$\|T_n\|_{[a,b]} \le \left(\frac{b-a}{4}\right)^{n-1} \quad \text{for } n = 1, 2, \dots$$

(see, for example, [48, Problem 15.9]).

Finally, for a hedgehog $K = K(\beta_1, \dots, \beta_d)$ with $\beta_j = \beta e^{2\pi i j/d}$, where $\beta > 0$, we have

$$\|P\|_K = \sup_{z\in K} |P(z)| = \sup_{z\in K} |P(\zeta^\ell z)| \quad \text{for } \ell \in \mathbb{Z}$$

implying

$$\|P\|_K^d = \sup_{z\in K} \left| \prod_{\ell=0}^{d-1} P(\zeta^\ell z) \right| = \sup_{z^d\in[0,\beta^d]} |\tilde{P}(z^d)| = \sup_{x\in[0,\beta^d]} |\tilde{P}(x)|.$$

Given $n = 1, 2, \dots$, the latter supremum is minimised by the monic polynomial $T_n(z)$ of degree n provided $\prod_{\ell=0}^{d-1} T_n(\zeta^\ell z) = \tilde{T}_n(z^d)$, where $\tilde{T}_n(x)$ is the nth Chebyshev polynomial on the interval $[0, \beta^d]$. Therefore,

$$t^*\big(K(\beta, \beta e^{2\pi i/d}, \dots, \beta e^{2\pi i(d-1)/d})\big)^d = t^*([0, \beta^d]) = \frac{\beta^d}{4}$$

so that $t^*\big(K(\beta, \beta e^{2\pi i/d}, \dots, \beta e^{2\pi i(d-1)/d})\big) = 4^{-1/d}\beta$.

For setting up some evidence towards Proposition 9.4, we first enlarge all the prickles to $[0, \beta_j'] \supset [0, \beta_j]$ of equal length $|\beta_1'| = \cdots = |\beta_d'| = \max_{1\le j\le d} |\beta_j|$. Since $K = K(\beta_1, \dots, \beta_d) \subset K(\beta_1', \dots, \beta_d') = K'$, we have $t(K) \le t(K')$ by the property of transfinite diameter. Therefore, it is sufficient to prove the statement for the case $|\beta_1| = \cdots = |\beta_d| = \beta$. Geometrically, the maximal possible value for all such configurations is achieved when the prickles are equidistributed around the origin, and in this case we have $t\big(K(\beta_1, \dots, \beta_d)\big) \le t^*\big(K(\beta_1, \dots, \beta_d)\big) = 4^{-1/d}\beta$ by the calculation above.

Chapter notes

The conjecture of Schinzel and Zassenhaus [69] was always in a shadow of Lehmer's question [18], about the infimum of the Mahler measure of a *monic* (non-cyclotomic) $P(x) = \prod_{j=1}^d (x-\alpha_j) \in \mathbb{Z}[x]$ (or of its zero $\alpha = \alpha_1$),

$$M(\alpha) = M(P(x)) = \prod_{j=1}^d \max\{1, |\alpha_j|\}.$$

All known lower bounds for $\overline{|\alpha|}$ were coming from those for $M(\alpha)$, and it was not even clear that a separate treatment of the former is possible. This makes Dimitrov's proof exceeding all expectations.

There is one more equivalent condition (iii) that can be included in Exercise 9.2:

$$\exp\bigg(\sum_{m=1}^\infty \frac{a_m x^m}{m}\bigg) \in \mathbb{Z}[[x]].$$

Its proof requires special tools (combinatorial, arithmetic or p-adic), which we do not touch here; the reader is advised to consult with the solution of Exercise 5.2 in [73, Chapter 5] for this.

The toughest ingredient of the proof is Dubinin's result (Proposition 9.4), for which a simple argument is not known. Some related discussions in this direction can be found in [46].

Chapter 10

Creative microscoping

In this chapter we return to the theme started in Chapter 1 — the q-deformation — with the motive to apply it analytically to proving congruences for integer and rational numbers. Such congruences clearly belong to arithmetic, so that we indeed witness another use of analysis in number theory.

We already know what q-numbers are and what q-factorials and q-binomials are. But we have not seen a q-version of the binomial theorem (1.2).

To feel ourselves comfortable about the material in this chapter we need to introduce relevant notation, which is in line with one from Chapters 3 and 7. The variable q will be treated either as a formal parameter or as a complex number inside the unit disk. For $m = 0, 1, \dots$, define first the q-shifted factorial

$$(a;q)_m = \prod_{j=1}^{m}(1 - aq^{j-1}),$$

also known as the q-Pochhammer symbol. It is not straightforward to observe its similarity with the Pochhammer symbol (3.15); in fact,

$$\lim_{q\to1}\frac{(q^s;q)_m}{(1-q)^m} = \lim_{q\to1}\prod_{j=1}^{m}[s+j-1]_q = \prod_{j=1}^{m}(s+j-1) = (s)_m.$$

When $|q| < 1$, the q-Pochhammer symbol makes perfect sense even if $m = \infty$ (something inaccessible to the usual Pochhammer symbol!).

Exercise 10.1 (q-binomial theorem). (a) Prove that for $n = 0, 1, 2, \dots$,

$$(x;q)_n = \sum_{m=0}^{n}\begin{bmatrix} n \\ m \end{bmatrix}_q (-1)^m q^{\binom{m}{2}} x^m. \tag{10.1}$$

(b) Verify that the limiting case of (10.1) as $q \to 1$ is $(1-x)^n = \sum_{m=0}^{n} \binom{n}{m}(-1)^m x^m$, which is an equivalent form of the binomial theorem (1.2).

In Section 10.3 we review further examples of such identities.

10.1 Supercongruences for binomial coefficients

Wilson's theorem (see Exercise 5.3) implies that for a prime p and a positive integer a we have $\binom{ap-1}{p-1} \equiv 1 \pmod{p}$, which can be stated equivalently as

$$\binom{ap}{p} \equiv \binom{a}{1} \pmod{p}.$$

It turns out that there is a much finer version of this result, which we discuss below.

The following congruence is usually attributed to Ljunggren (1952) or to Kazandzidis (1968), though it is essentially equivalent to its particular instance $a = 2$, $b = 1$ shown much earlier by Wolstenholme (1862).

Theorem 10.1. *Take $a > b > 0$ integers. Then for primes $p \geq 5$,*

$$\binom{ap}{bp} \equiv \binom{a}{b} \pmod{p^3}.$$

The term *supercongruence* is coined by Stienstra and Beukers to a congruence like in Theorem 10.1 when there is an 'unexpectedly' high power of p modulo which it takes place. At the same time the congruence has a relatively simple (or elementary) proof modulo p.

Instead of showing the Wolstenholme–Ljunggren–Kazandzidis supercongruence we will prove its q-deformed version. This is settled recently by Straub [74].

Theorem 10.2. *Take $a > b > 0$ integers. Then for integers $n > 0$,*

$$\begin{bmatrix} an \\ bn \end{bmatrix}_q \equiv \begin{bmatrix} a \\ b \end{bmatrix}_{q^{n^2}} - b(a-b)\binom{a}{b}\frac{n^2-1}{24}(q^n-1)^2 \pmod{\Phi_n(q)^3}, \qquad (10.2)$$

where $\Phi_n(q)$ denotes the nth cyclotomic polynomial.

Modulo $\Phi_n(q)^2$ rather than $\Phi_n(q)^3$ one can write down simpler versions, for example

$$\begin{bmatrix} an \\ bn \end{bmatrix}\sigma_n^b q^{\binom{bn}{2}} \equiv \binom{a-1}{b} + \binom{a-1}{a-b}\sigma_n^a q^{\binom{an}{2}} \pmod{\Phi_n(q)^2}, \qquad (10.3)$$

where

$$\sigma_n = (-1)^{n-1}.$$

The original proof given in [74] is combinatorial; here we follow a different route. The congruence in (10.2) is in fact a q-congruence, so that we have to clarify its meaning. A congruence $A_1(q) \equiv A_2(q) \pmod{P(q)}$ for rational functions $A_1(q), A_2(q)$ of parameters q and a polynomial $P(q) \in \mathbb{Z}[q]$ is understood as follows: the polynomial $P(q)$ is relatively prime with the denominators of $A_1(q)$ and $A_2(q)$, and $P(q)$ divides the numerator $A(q)$ of the difference $A_1(q) - A_2(q)$. The latter is equivalent to the condition that for each zero $\alpha \in \mathbb{C}$ of $P(q)$ of multiplicity k, the polynomial $(q-\alpha)^k$ divides $A(q)$ in $\mathbb{C}[q]$; in other words, $A_1(q) - A_2(q) = O\big((q-\alpha)^k\big)$ as $q \to \alpha$. This latter — purely analytic — interpretation underlies our argument in establishing q-congruences. For example, showing the congruence (10.3) is equivalent to verifying that

$$\begin{bmatrix} an \\ bn \end{bmatrix}_q (1-\varepsilon)^{\binom{bn}{2}} = \binom{a-1}{b} + \binom{a-1}{a-b}\sigma_n^a (1-\varepsilon)^{\binom{an}{2}} + O(\varepsilon^2) \quad \text{as } \varepsilon \to 0^+, \tag{10.4}$$

when $q = \zeta(1-\varepsilon)$ and ζ is any primitive nth root of unity.

How does Theorem 10.2 imply Theorem 10.1? The congruence (10.2) means that

$$\begin{bmatrix} an \\ bn \end{bmatrix}_q - \left(\begin{bmatrix} a \\ b \end{bmatrix}_{q^{n^2}} - b(a-b)\binom{a}{b}\frac{n^2-1}{24}(q^n-1)^2 \right) = \frac{1}{24}B(q)\Phi_n(q)^3$$

for some polynomial $B(q)$ with *integer* coefficients. Choosing $n = p > 3$ in this equality and then letting $q \to 1$ result in

$$\binom{ap}{bp} - \binom{a}{b} = \frac{1}{24}B_0 p^3 \quad \text{for some } B_0 \in \mathbb{Z},$$

so that Theorem 10.1 follows. Also notice that (10.3) simplifies to $\begin{bmatrix} an \\ bn \end{bmatrix}_q \equiv \begin{bmatrix} a \\ b \end{bmatrix}_{q^{n^2}}$ modulo $\Phi_n(q)^2$ (the additional term drops!), hence the above argument reduces the resulting congruence to

$$\binom{ap}{bp} \equiv \binom{a}{b} \pmod{p^2} \quad \text{for all primes } p,$$

the result first shown by Babbage (1819) for $a = 2$, $b = 1$ and preceding Wolstenholme's theorem.

Lemma 10.1. *Let ζ be a primitive nth root of unity. Then, as $q = \zeta(1-\varepsilon) \to \zeta$ radially,*

$$\begin{bmatrix} an \\ bn \end{bmatrix}_q \sigma_n^b q^{\binom{bn}{2}} - \binom{a-1}{b} - \binom{a-1}{a-b} \sigma_n^a q^{\binom{an}{2}}$$
$$= -b(a-b)\binom{a}{b} \frac{(3(an-1)^2 - an^2 - 1)n^2}{24} \varepsilon^2 + O(\varepsilon^3). \qquad (10.5)$$

Proof. It follows from the q-binomial theorem (10.1) with n replaced by an that

$$\frac{1}{n}\sum_{j=1}^{n} (\zeta^j x; q)_{an} = \sum_{\substack{m=0 \\ n|m}}^{an} \begin{bmatrix} an \\ m \end{bmatrix} (-x)^m q^{m(m-1)/2} = \sum_{b=0}^{a} \begin{bmatrix} an \\ bn \end{bmatrix} (-x)^{bn} q^{bn(bn-1)/2}. \qquad (10.6)$$

When $q = \zeta(1-\varepsilon)$, we get $\mathrm{d}/\mathrm{d}\varepsilon = -\zeta\,(\mathrm{d}/\mathrm{d}q)$. If

$$f(q) = (x;q)_{an} \quad \text{and} \quad g(q) = \frac{\mathrm{d}}{\mathrm{d}q} \log f(q) = -\sum_{\ell=1}^{an-1} \frac{\ell q^{\ell-1} x}{1 - q^\ell x},$$

then $f(q)|_{\varepsilon=0} = (1-x^n)^a$ and

$$\frac{\mathrm{d}f}{\mathrm{d}q} = fg, \quad \frac{\mathrm{d}^2 f}{\mathrm{d}q^2} = f\left(g^2 + \frac{\mathrm{d}g}{\mathrm{d}q}\right).$$

In particular,

$$\left.\frac{\mathrm{d}f}{\mathrm{d}\varepsilon}\right|_{\varepsilon=0} = (1-x^n)^a \sum_{\ell=1}^{an-1} \frac{\ell \zeta^\ell x}{1-\zeta^\ell x}$$

and

$$\left.\frac{\mathrm{d}^2 f}{\mathrm{d}\varepsilon^2}\right|_{\varepsilon=0} = (1-x^n)^a \left(\left(\sum_{\ell=1}^{an-1} \frac{\ell\zeta^\ell x}{1-\zeta^\ell x} \right)^2 - \sum_{\ell=1}^{an-1} \left(\frac{\ell^2 \zeta^{2\ell} x^2}{(1-\zeta^\ell x)^2} + \frac{\ell(\ell-1)\zeta^\ell x}{1-\zeta^\ell x} \right) \right).$$

Further observe the following summation formulae:

$$\frac{1}{n}\sum_{j=1}^{n} \left.\frac{x}{1-x}\right|_{x \mapsto \zeta^j x} = \frac{x^n}{1-x^n},$$

$$\frac{1}{n}\sum_{j=1}^{n} \left.\left(\frac{x}{1-x}\right)^2\right|_{x \mapsto \zeta^j x} = \frac{nx^n}{(1-x^n)^2} - \frac{x^n}{1-x^n}$$

and

$$\frac{1}{n}\sum_{j=1}^{n}\frac{x}{1-x}\,\frac{\zeta^k x}{1-\zeta^k x}\bigg|_{x\mapsto\zeta^j x}=-\frac{x^n}{1-x^n}\quad\text{for } k\not\equiv 0 \pmod n.$$

Implementing this information into (10.6) we obtain

$$\begin{aligned}\sum_{b=0}^{a}\begin{bmatrix}an\\bn\end{bmatrix}(-x)^{bn}q^{bn(bn-1)/2}\bigg|_{q=\zeta(1-\varepsilon)}&=(1-x^n)^a\Bigg(1+\varepsilon\,\frac{x^n}{1-x^n}\sum_{\ell=1}^{an-1}\ell\\&-\frac{\varepsilon^2}{2}\,\frac{x^n}{1-x^n}\Bigg(\sum_{\ell=1}^{an-1}\ell\Bigg)^2+\frac{\varepsilon^2}{2}\,\frac{nx^n}{(1-x^n)^2}\sum_{\substack{\ell_1,\ell_2=1\\ \ell_1\equiv\ell_2 \ (\mathrm{mod}\ n)}}^{an-1}\ell_1\ell_2\\&-\frac{\varepsilon^2}{2}\bigg(\frac{nx^n}{(1-x^n)^2}-\frac{x^n}{1-x^n}\bigg)\sum_{\ell=1}^{an-1}\ell^2-\frac{\varepsilon^2}{2}\,\frac{x^n}{1-x^n}\sum_{\ell=1}^{an-1}\ell(\ell-1)\Bigg)+O(\varepsilon^3).\end{aligned}$$

Finally, compare the coefficients of powers of x^n on both sides of the relation obtained; this way we arrive at the asymptotics in (10.5). □

To prove Theorem 10.2 we need to produce a 'matching' asymptotic for

$$\begin{bmatrix}a\\b\end{bmatrix}_{q^{n^2}}.$$

This happens to be easier than what we have done in Lemma 10.1, because $q^{n^2}=(1-\varepsilon)^{n^2}$ does not depend on the choice of primitive nth root of unity ζ when $q=\zeta(1-\varepsilon)$.

Lemma 10.2. *As $q=\zeta(1-\varepsilon)\to\zeta$ radially,*

$$\begin{aligned}&\begin{bmatrix}a\\b\end{bmatrix}_{q^{n^2}}\sigma_n^b q^{\binom{bn}{2}}-\binom{a-1}{b}-\binom{a-1}{a-b}\sigma_n^a q^{\binom{an}{2}}\\&\qquad=-b(a-b)\binom{a}{b}\frac{(3(an-1)^2-(a+1)n^2)n^2}{24}\,\varepsilon^2+O(\varepsilon^3).\end{aligned}$$

Proof. From (10.1) we conclude that

$$(x^n q^{\binom{n}{2}};q^{n^2})_a=\sum_{b=0}^{a}\begin{bmatrix}a\\b\end{bmatrix}_{q^{n^2}}\sigma_n^b(-x)^{bn}q^{\binom{bn}{2}}.$$

Then, for $q=\zeta(1-\varepsilon)$, we write $y=\sigma_n x^n$ to obtain

$$\begin{aligned}(x^n q^{\binom{n}{2}};q^{n^2})_a&=(y(1-\varepsilon)^{\binom{n}{2}};(\varepsilon)^{n^2})_a=\prod_{\ell=0}^{a-1}\big(1-y(1-\varepsilon)^{\ell n^2+\binom{n}{2}}\big)\\&=(1-y)^a\prod_{\ell=0}^{a-1}\Bigg(1-\frac{y}{1-y}\sum_{i=1}^{\ell n^2+\binom{n}{2}}\binom{\ell n^2+\binom{n}{2}}{i}(-\varepsilon)^i\Bigg).\end{aligned}$$

It remains to compare the coefficients of x^n on both sides. □

Proof of Theorem 10.2. Note that $\varepsilon = \frac{1}{n}(1-q^n)+O(\varepsilon^2)$ as $q=\zeta(1-\varepsilon)\to\zeta$ radially, where ζ is primitive nth root of unity. Combining the expansions in Lemmas 10.1, 10.2 we find out that

$$\begin{aligned}
&\begin{bmatrix} an \\ bn \end{bmatrix}_q \sigma_n^b q^{\binom{bn}{2}} - \begin{bmatrix} a \\ b \end{bmatrix}_{q^{n^2}} \sigma_n^b q^{\binom{bn}{2}} \\
&\quad = b(a-b)\binom{a}{b}\frac{(an^2+1)n^2-(a+1)n^4}{24}\,\varepsilon^2+O(\varepsilon^3) \\
&\quad = -b(a-b)\binom{a}{b}\frac{(n^2-1)n^2}{24}\,\varepsilon^2+O(\varepsilon^3) \\
&\quad = -b(a-b)\binom{a}{b}\frac{n^2-1}{24}\,(q^n-1)^2+O(\varepsilon^3).
\end{aligned}$$

This means that the difference of both sides is divisible by $(q-\zeta)^3$ for any nth primitive root of unity ζ, hence by $\Phi_n(q)^3$. The latter property is equivalent to the congruence (10.2). □

10.2 Ramanujan's formulae for $1/\pi$

Srinivasa Ramanujan (1887–1920) was an Indian mathematician whose mathematical contributions had a lasting impact on the development of number theory and special functions. Many notions and theorems originated from his papers, letters and notebooks; the account of his work and its implications can be found in [2, 10, 11, 60].

In his development of the theory of elliptic functions, Ramanujan came up [65] with computationally efficient representations of $1/\pi$. Examples are

$$\sum_{n=0}^{\infty}\frac{(\frac12)_n^3}{n!^3}\,(1+6n)\,\frac{1}{2^{2n}}=\frac{4}{\pi}, \tag{10.7}$$

$$\sum_{n=0}^{\infty}\frac{(\frac12)_n(\frac13)_n(\frac23)_n}{n!^3}\,(4+33n)\,\frac{2^{2n}}{5^{3n}}=\frac{15\sqrt{3}}{2\pi}, \tag{10.8}$$

$$\sum_{n=0}^{\infty}\frac{(\frac12)_n(\frac16)_n(\frac56)_n}{n!^3}\,(8+133n)\left(\frac{4}{85}\right)^{3n}=\frac{85\sqrt{255}}{54\pi}, \tag{10.9}$$

$$\sum_{n=0}^{\infty}\frac{(\frac12)_n(\frac14)_n(\frac34)_n}{n!^3}\,(1123+21460n)\left(-\frac{1}{882^2}\right)^{n}=\frac{4\cdot 882}{\pi}, \tag{10.10}$$

$$\sum_{n=0}^{\infty}\frac{(\frac12)_n(\frac14)_n(\frac34)_n}{n!^3}\,(1103+26390n)\,\frac{1}{99^{4n}}=\frac{99^2}{2\pi\sqrt{2}}, \tag{10.11}$$

where $(s)_n = \prod_{j=1}^{n}(s+j-1)$ is the Pochhammer symbol (3.15); these are equations (28), (32), (34), (39) and (44) on the list in [65]. Ramanujan did not hide his interest in computing π; his comment in [65] about identity (10.11) says "The last series (44) is extremely rapidly convergent." In total, Ramanujan gave seventeen such equalities.

The identities do not look hard. In spite of this, their first proofs were only obtained in the 1980s by the Borweins and independently by the Chudnovskys. A historical account of contemporary techniques for proving Ramanujan's (and Ramanujan-type) formulae for $1/\pi$ can be found in [8, 88]. The dominating method which, for example, works for any formulae in (10.7)–(10.11) is based on modular-function parametrisations of the underlying series. This modular technique cannot be counted as elementary, but it leads to many further examples (though not necessarily computationally useful) like the formula

$$\sum_{n=0}^{\infty} u_n \cdot (20n + 10 - 3\sqrt{5})\left(\frac{\sqrt{5}-1}{2}\right)^{12n} = \frac{20\sqrt{3}+9\sqrt{15}}{6\pi} \tag{10.12}$$

of Ramanujan type, involving the Apéry numbers

$$u_n = \sum_{k=0}^{n} \binom{n+k}{n}^2 \binom{n}{k}^2 \tag{10.13}$$

from Section 7.1; this identity was discovered by T. Sato in 2002.

One formula, which is not on Ramanujan's list in [65] but clearly belongs to it, is

$$\sum_{n=0}^{\infty} \frac{(\frac{1}{2})_n^3}{n!^3}(1+4n)\cdot(-1)^n = \frac{2}{\pi}. \tag{10.14}$$

In fact, this identity was proven by Bauer (1859) long before Ramanujan was born, using a quite elementary argument. The convergence in (10.14) is poor and comparable with Leibniz's formula (1.11) (though the latter is for π itself). Nevertheless the shape of the formula is very much the same as in (10.7)–(10.11), with the sums on left-hand sides are linked with some particular instances $m = 3$ of the (generalized) hypergeometric series

$${}_mF_{m-1}\left(\begin{matrix} a_1, a_2, \dots, a_m \\ b_2, \dots, b_m \end{matrix} \,\middle|\, z\right) = \sum_{n=0}^{\infty} \frac{(a_1)_n (a_2)_n \cdots (a_m)_n}{(b_2)_n \cdots (b_m)_n} \frac{z^n}{n!}. \tag{10.15}$$

Namely, the identities listed all involve linear combinations of ${}_3F_2$ series and its derivative at a (rational) point, with $a_1 = \frac{1}{2}$, $a_2 = 1 - a_3 \in \{\frac{1}{2}, \frac{1}{3}, \frac{1}{4}, \frac{1}{6}\}$ and $b_2 = b_3 = 1$. The series defining ${}_mF_{m-1}(z)$ converges in the unit disk

$|z| < 1$. It also satisfies a linear homogeneous differential equation of order m with coefficients in $\mathbb{C}(z)$; this differential equation allows one to continue the function analytically to $\mathbb{C} \setminus [1, +\infty)$. Good sources for the theory of generalized hypergeometric series are books [6, 82] (see also [71]). Though (10.12) does not belongs to this hypergeometric family of Ramanujan's formulae, the generating series $\sum_{n=0}^{\infty} u_n z^n$ satisfies a third order linear differential equation (given in an equivalent form in Exercise 7.2) which shares many similarities with those satisfied by ${}_3F_2(z)$; this places (10.12) on the list of Ramanujan-type formulae.

The following exercise illustrates another technique which can be used for proving *some* identities of Ramanujan type. It relies on the method of creative telescoping which we have already seen in action in Exercise 7.2.

Exercise 10.2 (Zeilberger [28]). Define

$$F(n,k) = \frac{(\frac{1}{2})_n^2(-k)_n}{n!^2(\frac{3}{2}+k)_n}\,(1+4n)\,(-1)^n \cdot \frac{\Gamma(\frac{3}{2})\Gamma(1+k)}{\Gamma(\frac{3}{2}+k)}$$

and take

$$G(n,k) = \frac{(2n+1)^2}{(2n+2k+3)(4n+1)}\,F(n,k).$$

(a) Show that for $n = 0, 1, 2, \ldots$ and $k = 0, 1, 2 \ldots$,

$$F(n,k+1) - F(n,k) = G(n,k) - G(n-1,k).$$

(b) Use part (a) to prove that

$$\sum_{n=0}^{\infty} F(n,k) = \sum_{n\in\mathbb{Z}} F(n,k)$$

does not depend on k. Then show that this constant is 1 (computing the sum, for example, at $k = 0$).

(c) Conclude that

$$\sum_{n=0}^{\infty} \frac{(\frac{1}{2})_n^2(-k)_n}{n!^2(\frac{3}{2}+k)_n}\,(1+4n)(-1)^n = \frac{\Gamma(\frac{3}{2}+k)}{\Gamma(\frac{3}{2})\Gamma(1+k)}. \tag{10.16}$$

Hint. (a) Divide both sides by $F(n,k)$ to reduce verification to one of an identity for simple rational functions in n and k. □

Though equality (10.16) is only shown to be true for $k = 0, 1, 2, \ldots$, it remains true for $k \in \mathbb{C}$ with $\operatorname{Re} k > -1$ — this is a consequence of Carlson's theorem (see, for example, [6, Section 5.3]), another classical analysis result. Finally, notice that Bauer's identity (10.14) is the case $k = -1/2$ of (10.16).

Another elementary technique for producing new Ramanujan-type identities from already known ones is known as the translation method [23, 42]. It sources from algebraic identities of hypergeometric series and relies on manipulations that use calculus rules. It is illustrated in the following exercise.

Exercise 10.3. (a) Show Bailey's cubic transformation

$$\sum_{n=0}^{\infty} \frac{(\frac{1}{2})_n^3}{n!^3} x^n = (1-4x)^{-1/2} \sum_{n=0}^{\infty} \frac{(\frac{1}{2})_n(\frac{1}{6})_n(\frac{5}{6})_n}{n!^3} \left(-\frac{27x}{(1-4x)^3}\right)^n$$

for x from a neighbourhood of the origin.

(b) Using the identity from part (a) and its x-derivative at $x = -1$, show that

$$\sum_{n=0}^{\infty} \frac{(\frac{1}{2})_n(\frac{1}{6})_n(\frac{5}{6})_n}{n!^3} (3+28n)\left(\frac{3}{5}\right)^{3n} = \frac{5\sqrt{5}}{\pi}.$$

This formula was not given by Ramanujan in [65].

Hint. (a) Verify that both sides satisfy the same linear differential equation of order 3.

(b) Apply the operator $\mathrm{Id} + 4x\frac{\mathrm{d}}{\mathrm{d}x}$ to both sides of identity from part (a), then substitute $x = -1$ and use the known formula (10.14) for the left-hand side. □

The next example is an advanced version of Exercise 10.3.

Exercise 10.4. Let u_n be the sequence of Apéry numbers defined in (10.13).

(a) Show that for sufficiently small $|x|$,

$$\sum_{n=0}^{\infty} u_n \frac{x^n(1-8x)^n}{(1+x)^{n+1}} = (1-8x)^{-3/2} \sum_{n=0}^{\infty} \frac{(\frac{1}{2})_n^3}{n!^3} \left(-\frac{64x(1+x)^3}{(1-8x)^3}\right)^n.$$

(b) Use the transformation from part (a) at $x = (9\sqrt{6}-22)/4$ and (10.14) to prove

$$\sum_{n=0}^{\infty} (4-\sqrt{6}+8n)u_n(\sqrt{3}-\sqrt{2})^{4n+2} = \frac{1}{\pi\sqrt{2}}.$$

In 1997 Van Hamme noticed that several formulae of Ramanujan for infinite sums possess arithmetic finite-sum analogues. The example relevant to our discussion in this section and corresponding to Ramanujan-type formula (10.14) is the family of congruences

$$\sum_{n=0}^{p-1} \frac{(\frac{1}{2})_n^3}{n!^3}(1+4n)(-1)^n \equiv \left(\frac{-1}{p}\right)p \pmod{p^3} \quad \text{for primes } p > 2, \qquad (10.17)$$

where $\left(\frac{a}{p}\right)$ denote the Legendre symbol (see Exercise 5.10). This was subsequently proven by Mortenson (2008) and several other proofs appeared later. It has been also realised [89] (mostly numerically!) that the pattern continues to hold for other Ramanujan's and Ramanujan-type formula (at least when they correspond to ${}_3F_2(z)$ series with rational z), so that we have

$$\sum_{n=0}^{p-1} \frac{(\frac{1}{2})_n^3}{n!^3} (1+6n) \frac{1}{2^{2n}} \equiv \left(\frac{-1}{p}\right) p \pmod{p^3} \quad \text{for } p>2, \tag{10.18}$$

and

$$\sum_{n=0}^{p-1} \frac{(\frac{1}{2})_n(\frac{1}{3})_n(\frac{2}{3})_n}{n!^3} (4+33n) \frac{2^{2n}}{5^{3n}} \equiv 4\left(\frac{-3}{p}\right) p \pmod{p^3} \quad \text{for } p>3,$$

$$\sum_{n=0}^{p-1} \frac{(\frac{1}{2})_n(\frac{1}{6})_n(\frac{5}{6})_n}{n!^3} (8+133n)\left(\frac{4}{85}\right)^{3n} \equiv 8\left(\frac{-255}{p}\right) p \pmod{p^3} \quad \text{for } p>5,\ p\ne 17,$$

$$\sum_{n=0}^{p-1} \frac{(\frac{1}{2})_n(\frac{1}{4})_n(\frac{3}{4})_n}{n!^3} (1123+21460n)\left(-\frac{1}{882^2}\right)^{n} \equiv 1123\left(\frac{-1}{p}\right) p \pmod{p^3} \quad \text{for } p>3,\ p\ne 7,$$

$$\sum_{n=0}^{p-1} \frac{(\frac{1}{2})_n(\frac{1}{4})_n(\frac{3}{4})_n}{n!^3} (1103+26390n) \frac{1}{99^{4n}} \equiv 1103\left(\frac{-2}{p}\right) p \pmod{p^3} \quad \text{for } p>3,\ p\ne 11,$$

as p-counterparts of (10.7)–(10.11). At the moment the general congruences from this list are only proven for the family (10.18).

Notice that the terms in these sums are not integers but rational numbers, however with the denominators that only involve finitely many (small) primes which we exclude from the consideration. To see that we just need to note that

$$\frac{(\frac{1}{2})_n^3}{n!^3} = 2^{-6n}\binom{2n}{n}^3, \qquad \frac{(\frac{1}{2})_n(\frac{1}{3})_n(\frac{2}{3})_n}{n!^3} = 2^{-2n}3^{-3n}\binom{2n}{n}\frac{(3n)!}{n!^3},$$

$$\frac{(\frac{1}{2})_n(\frac{1}{4})_n(\frac{3}{4})_n}{n!^3} = 2^{-8n}\frac{(4n)!}{n!^4}, \qquad \frac{(\frac{1}{2})_n(\frac{1}{6})_n(\frac{5}{6})_n}{n!^3} = 12^{-3n}\frac{(6n)!}{n!^3(3n)!},$$

where the factorial ratios are all integral.

In spite of their limited capacity, there are already several methods on the market designed for proving Ramanujan-type supercongruences. One method, which is based on ideas used in Section 10.1 and known as creative microscoping, makes more (when succeed): it leads to simultaneous proofs of Ramanujan-type identity and corresponding Ramanujan-type supercongruences, thus explaining their mysterious interconnection. In the rest of this chapter we illustrate the performance of creative microscoping on the pair (10.14), (10.17).

10.3 q-Hypergeometry

We have already witnessed in the proof of Lemma 10.1 a use of the q-binomial theorem (10.1). In fact, the latter formula comes as a particular case of a more general result.

Theorem 10.3 (q-binomial theorem). *When $|q| < 1$ and $|z| < 1$,*

$$\sum_{n=0}^{\infty} \frac{(a;q)_n}{(q;q)_n} z^n = \frac{(az;q)_\infty}{(z;q)_\infty}. \tag{10.19}$$

This theorem is a q-extension of the general binomial formula

$$(1-z)^{-a} = \sum_{n=0}^{\infty} \frac{(a)_n}{n!} z^n = {}_1F_0\left(\begin{matrix} a \\ \end{matrix} \,\middle|\, z\right)$$

(see (10.15)), and this extension is a fundamental identity in the theory of q-hypergeometric functions: it is expected that every other q-hypergeometric identity can be deduced via a finite combination of equation (10.19) (of course, with different setup for its parameters).

Proof. We follow the creative telescoping strategy. Denote the nth term of the sum in (10.19) by $F_n(z)$ and take

$$G_n = \frac{1-q^n}{z-1} F_n(z) = \begin{cases} \dfrac{(a;q)_n \, z^n}{(q;q)_{n-1} \, (z-1)} & \text{if } n > 0, \\ 0 & \text{if } n = 0. \end{cases}$$

We claim the telescoping relation

$$F_n(z) - \frac{1-az}{1-z} F_n(zq) = G_{n+1} - G_n$$

for $n = 0, 1, 2, \dots$; division of both sides by $F_n(z)$ reduces the equality to a simpler one,

$$1 - \frac{1-az}{1-z} X = \frac{(1-aX)z}{z-1} - \frac{1-X}{z-1} \quad \text{where } X = q^n,$$

whose verification is straightforward. Summing both sides of the telescoping relation over $n = 0, 1, 2, \dots$ results in

$$\sum_{n=0}^{\infty} F_n(z) - \frac{1-az}{1-z} \sum_{n=0}^{\infty} F_n(zq) = 0.$$

Iterating this equality m times leads to

$$\sum_{n=0}^{\infty} F_n(z) = \frac{1-az}{1-z} \sum_{n=0}^{\infty} F_n(zq) = \cdots = \frac{(az;q)_m}{(z;q)_m} \sum_{n=0}^{\infty} F_n(zq^m),$$

and (10.19) follows from taking the limit as $m \to \infty$ in the result. □

To see that (10.1) is a special case of Theorem 10.3, replace the summation index n in (10.19) by m and then take $a = q^{-n}$, $z = xq^n$.

One particular feature that makes the creative telescoping possible in the above proof but also for general (q-)hypergeometric sums $\sum_{n=0}^{\infty} c_n$ is a simple form of the quotient of two consecutive terms of the latter. This brings us naturally to a definition of (q-)hypergeometric series: it is one for which c_{n+1}/c_n is a rational function of index n (respectively, of parameter q^n). You may check that (10.15) is a hypergeometric series and that all the q-sums in this chapter are q-hypergeometric series.

It is absolutely amazing how rich a hierarchy of q-hypergeometric identities (summations and transformations) is. To get a good view of it one needs to master numerous available tools; a comprehensive source of those is the book [34] known among the specialists as the q-Bible. Below we limit ourselves to a particular q-hypergeometric summation, which is a fine representative of the theory and at the same time an instrument required in our arithmetic application. (In the q-Bible it is inelegantly called the summation formula for a non-terminating very-well-poised ${}_6\phi_5$-series; see [34, eq. (II.20)].)

Theorem 10.4. *When* $|q| < 1$ *and* $|aq| < |bcd|$,

$$\sum_{n=0}^{\infty} \frac{(1-aq^{2n})\,(a;q)_n(b;q)_n(c;q)_n(d;q)_n}{(1-a)\,(q;q)_n(aq/b;q)_n(aq/c;q)_n(aq/d;q)_n} \left(\frac{aq}{bcd}\right)^n$$
$$= \frac{(aq;q)_\infty(aq/(bc);q)_\infty(aq/(bd);q)_\infty(aq/(cd);q)_\infty}{(aq/b;q)_\infty(aq/c;q)_\infty(aq/d;q)_\infty(aq/(bcd);q)_\infty}. \tag{10.20}$$

Proof. Let $F_n(a)$ denote the nth term of the sum in (10.20). Then

$$F_n(a/q) - \frac{(a-1)(a-bc)(a-bd)(a-cd)}{(a-b)(a-c)(a-d)(a-bcd)} F_n(a) = G_{n+1} - G_n$$

for all $n = 0, 1, 2, \dots$, where

$$G_n = \frac{(a^2q^n - bcdq)(1 - q^n)}{(aq^{2n} - q)(a - bcd)} \cdot F_n(a/q),$$

in particular $G_0 = 0$. (As before, verification commences after reduction to a rational-function identity via division of both sides by $F_n(a/q)$.) Replacing a with aq, summing the telescoping relation over $n = 0, 1, 2, \dots$ and using $G_n \to 0$ as $n \to \infty$ when $|aq/(bcd)| < 1$ we obtain

$$\begin{aligned}\sum_{n=0}^{\infty} F_n(a) &= \frac{(aq-1)(aq-bc)(aq-bd)(aq-cd)}{(aq-b)(aq-c)(aq-d)(aq-bcd)} \sum_{n=0}^{\infty} F_n(aq) \\ &= \frac{(1-aq)(1-aq/(bc))(1-aq/(bd))(1-aq/(cd))}{(1-aq/b)(1-aq/c)(1-aq/d)(1-aq/(bcd))} \sum_{n=0}^{\infty} F_n(aq).\end{aligned}$$

It remains to iterate the result m times and then compute the limit as $m \to \infty$. □

10.4 Supercongruences and q-supercongruences

Recall the notation $[m] = [m]_q = (1 - q^m)/(1 - q)$ for the q-numbers.

Theorem 10.5 (q-analogue of equation (10.14)). *The following equality is true:*

$$\sum_{n=0}^{\infty} \frac{(q;q^2)_n^3}{(q^2;q^2)_n^3} [1+4n]_q \cdot (-1)^n q^{n^2} = \frac{(q^3;q^2)_\infty (q;q^2)_\infty}{(q^2;q^2)_\infty^2}. \qquad (10.21)$$

Theorem 10.6 (q-analogue of family (10.17)). *Let m be a positive odd integer. Then*

$$\sum_{n=0}^{m-1} \frac{(q;q^2)_n^3}{(q^2;q^2)_n^3} [1+4n] \cdot (-1)^n q^{n^2} \equiv q^{(m-1)^2/4} [m] \left(\frac{-1}{m}\right) \pmod{[m]\Phi_m(q)^2}. \qquad (10.22)$$

In the last theorem, the truncated q-hypergeometric sums are considered modulo (products of) cyclotomic polynomials. Notice that $[m]_q = \prod_{d|m,\, d>1} \Phi_d(q)$ and that $[p]_q = \Phi_p(q) \to p$ as $q \to 1$ when p is prime.

In the case of formula (10.21), we see that

$$\lim_{q\to 1} \frac{(q;q^2)_n}{(q^2;q^2)_n} = \frac{(\frac{1}{2})_n}{n!}$$

and

$$\lim_{q\to 1}\frac{(q;q^2)_\infty}{(q^2;q^2)_\infty(1-q^2)^{1/2}}=\frac{1}{\Gamma(\frac{1}{2})}=\frac{1}{\sqrt{\pi}},$$

hence in the limit as $q\to 1$ we obtain (10.14). At the same time, taking the limit as $q\to 1$ in (10.22) for $m=p$ prime leads to the Ramanujan-type supercongruences (10.17).

Our proof of Theorem 10.6 combines two principles. One corresponds to achieving the congruences in (10.22) modulo $[m]$ only, and for this we deal with the q-hypergeometric sum (10.21) at a 'q-microscopic' level — that is, at roots of unity (and this cannot be transformed into a derivation of (10.17) directly from (10.14)). Another 'creative' principle is about getting more parameters involved in the q-story.

Theorem 10.7. *Let m be a positive odd integer. Then, for any indeterminates a and q, we have modulo $[m](1-aq^m)(a-q^m)$,*

$$\sum_{n=0}^{m-1}\frac{(q;q^2)_n(aq;q^2)_n(q/a;q^2)_n}{(q^2;q^2)_n(aq^2;q^2)_n(q^2/a;q^2)_n}[1+4n]\,(-1)^nq^{n^2}\equiv q^{(m-1)^2/4}[m]\left(\frac{-1}{m}\right). \tag{10.23}$$

Proof of Theorem 10.6. The denominator of (10.23) related to a is the factor $(aq^2;q^2)_{m-1}(q^2/a;q^2)_{m-1}$; its limit as $a\to 1$ is relatively prime to $\Phi_m(q)$, since m is odd. On the other hand, the limit of $(1-aq^m)(a-q^m)$ as $a\to 1$ has the factor $\Phi_m(q)^2$. Thus, letting $a\to 1$ in (10.23) we see that (10.22) is true modulo $\Phi_m(q)^3$. At the same time, by considering (10.23) modulo $[m]$ only and specialising $a=1$ in the result reads

$$\sum_{n=0}^{m-1}\frac{(q;q^2)_n(aq;q^2)_n(q/a;q^2)_n}{(q^2;q^2)_n(aq^2;q^2)_n(q^2/a;q^2)_n}[1+4n]\,(-1)^nq^{n^2}\equiv 0 \pmod{[m]}.$$

Thus, indeed both sides of (10.22) are congruent modulo $[m]\Phi_m(q)^2$. □

In turn, the general set of congruences in Theorem 10.7 is deduced from a non-terminating version of (10.23).

Theorem 10.8. *The following identity is true:*

$$\sum_{n=0}^{\infty}\frac{(1-q^{4n+1})\,(q;q^2)_n(aq;q^2)_n(q/a;q^2)_n}{(1-q)\,(q^2;q^2)_n(aq^2;q^2)_n(q^2/a;q^2)_n}(-1)^nq^{n^2}$$
$$=\frac{(q^3;q^2)_\infty(q;q^2)_\infty}{(aq^2;q^2)_\infty(q^2/a;q^2)_\infty}. \tag{10.24}$$

Proof. Take $d = 1/\varepsilon$ in (10.20) and let $\varepsilon \to 0$ to obtain

$$\sum_{n=0}^{\infty} \frac{(1-aq^{2n})\,(a;q)_n(b;q)_n(c;q)_n(-1)^n q^{n(n-1)/2}}{(1-a)\,(q;q)_n(aq/b;q)_n(aq/c;q)_n}\left(\frac{aq}{bc}\right)^n$$
$$= \frac{(aq;q)_\infty(aq/(bc);q)_\infty}{(aq/b;q)_\infty(aq/c;q)_\infty}.$$

In this identity replace q with q^2, then choose $a = q$, $b = dq$, $c = q/d$ and finally replace d with a. □

Proof of Theorem 10.5. Take $a = 1$ in (10.24). □

In the remainder of this section we discuss the most non-trivial part of the method of creative microscoping — deduction of Theorem 10.7 from Theorem 10.8.

Lemma 10.3. *Let m be a positive odd integer. Then*

$$\sum_{n=0}^{m-1} \frac{(q;q^2)_n(q^{1-m};q^2)_n(q^{1+m};q^2)_n}{(1-q)\,(q^2;q^2)_n(q^{2-m};q^2)_n(q^{2+m};q^2)_n}\,[1+4n]\,(-1)^n q^{n^2}$$
$$= q^{(m-1)^2/4}[m]\left(\frac{-1}{m}\right). \tag{10.25}$$

Proof. We substitute $a = q^m$ into (10.24). Then the left-hand side of (10.24) terminates (already at $n = (m-1)/2$, meaning that all its terms starting from $(m+1)/2$ vanish) and equals the sum in (10.25). On the other hand, the substitution transforms the right-hand side of (10.24) into

$$\frac{(q^3;q^2)_\infty(q;q^2)_\infty}{(q^{2-m};q^2)_\infty(q^{2+m};q^2)_\infty} = \frac{(q^3;q^2)_{(m-1)/2}}{(q^{2-m};q^2)_{(m-1)/2}}$$
$$= \frac{(q^3;q^2)_{(m-1)/2}}{(-1)^{(m-1)/2}q^{-(m-1)^2/4}(q;q^2)_{(m-1)/2}}$$
$$= (-1)^{(m-1)/2}q^{(m-1)^2/4}[m].$$ □

Proof of Theorem 10.7. Let $\zeta \ne 1$ be a primitive dth root of unity, where $d \mid m$ and $m > 1$ is odd (hence d is odd as well). Denote by

$$F_n(q) = \frac{(q;q^2)_n(aq;q^2)_n(q/a;q^2)_n}{(q^2;q^2)_n(aq^2;q^2)_n(q^2/a;q^2)_n}\,[1+4n]\,(-1)^n q^{n^2}$$

the nth term of the sum (10.24) and write (10.24) as

$$\sum_{\ell=0}^{\infty} F_{\ell d}(q) \sum_{n=0}^{d-1} \frac{F_{\ell d+n}(q)}{F_{\ell d}(q)} = \frac{(q^3;q^2)_\infty(q;q^2)_\infty}{(aq^2;q^2)_\infty(q^2/a;q^2)_\infty}. \tag{10.26}$$

Consider the limit as $q \to \zeta$ radially, that is, $q = r\zeta$ where $r \to 1^-$. On the left-hand side we get

$$\lim_{q\to\zeta} \frac{F_{\ell d+n}(q)}{F_{\ell d}(q)} = \frac{F_{\ell d+n}(\zeta)}{F_{\ell d}(\zeta)} = F_n(\zeta)$$

and

$$\lim_{q\to\zeta} F_{\ell d}(q) = \lim_{q\to\zeta} \frac{(q;\zeta^2)_n(aq;\zeta^2)_n(q/a;\zeta^2)_n}{(q^2;\zeta^2)_n(aq^2;\zeta^2)_n(q^2/a;\zeta^2)_n}(-1)^\ell = (-1)^\ell,$$

since d is odd and $(a;\zeta^2)_{\ell d} = (a;\zeta^2)_d^\ell = (1-a^d)^\ell$. For the right-hand side of (10.26),

$$\lim_{q\to\zeta} \frac{(q^3;q^2)_\infty(q;q^2)_\infty}{(aq^2;q^2)_\infty(q^2/a;q^2)_\infty} = 0,$$

because the part $(q;q^2)_{(d+1)/2}$ of the product $(q;q^2)_\infty$ vanishes at $q = \zeta$. By comparing the asymptotics of both sides of (10.26) as $q \to \zeta$ we conclude that

$$\sum_{n=0}^{d-1} F_n(\zeta) = 0;$$

this in turn implies that

$$\sum_{n=0}^{m-1} F_n(\zeta) = \sum_{n=0}^{d-1} F_n(\zeta) + \sum_{n=d}^{2d-1} F_n(\zeta) + \cdots + \sum_{n=m-d}^{m-1} F_n(\zeta) = \frac{n}{d}\sum_{n=0}^{d-1} F_n(\zeta) = 0.$$

Since this is true for any choice of dth root of unity ζ, the equality can be stated as the congruence $\sum_{n=0}^{m-1} F_n(q) \equiv 0 \pmod{\Phi_d(q)}$. The latter is valid for any $d \mid m$, $d > 1$, hence

$$\sum_{n=0}^{m-1} F_n(q) \equiv 0 \equiv q^{(m-1)^2/4}[m]\left(\frac{-1}{m}\right) \pmod{[m]}.$$

On the other hand, it follows from Lemma 10.3 that

$$\sum_{n=0}^{m-1} \frac{(q;q^2)_n(aq;q^2)_n(q/a;q^2)_n}{(q^2;q^2)_n(aq^2;q^2)_n(q^2/a;q^2)_n}[1+4n](-1)^n q^{n^2} = q^{(m-1)^2/4}[m]\left(\frac{-1}{m}\right)$$

when $a = q^m$ or $a = q^{-m}$; this implies that the congruences (10.23) hold true modulo $1 - aq^m$ and $a - q^m$. Since the polynomials $[m]$, $1 - aq^m$ and $a - q^m$ are relatively prime, we obtain (10.23) modulo their product. □

We can summarise our derivation path of the results as follows:

$$\begin{array}{ccccc}
\text{Theorem 10.8} & \underset{a=1}{\Longrightarrow} & \text{Theorem 10.5} & \underset{q\to1}{\Longrightarrow} & \text{formula (10.14)} \\
\Downarrow & & & & \\
\text{Theorem 10.7} & \underset{a\to1}{\Longrightarrow} & \text{Theorem 10.6} & \underset{q\to1}{\Longrightarrow} & \text{congruences (10.17)}
\end{array}$$

The top of this scheme — Theorem 10.8 — comes essentially for free from a known q-hypergeometric identity, and many further entries from [34] lead to remarkable (and quite difficult!) congruences, so that the q-Bible turns out to be a treasury book for number theory.

Chapter notes

There is a modulo p^4 extension of Theorem 10.1,

$$\binom{ap}{bp} \equiv \binom{a}{b} + ab(a-b)\binom{a}{b}p\sum_{k=1}^{p-1}\frac{1}{k} \pmod{p^4} \quad \text{for prime } p > 3.$$

It involves the harmonic sums

$$\sum_{k=1}^{p-1}\frac{1}{k} \equiv 0 \pmod{p^2} \quad \text{for prime } p > 2,$$

and can be also deduced from suitable q-extensions using the method in Section 10.1.

The theme of Ramanujan-type formulae for $1/\pi$ is quite rich, we do not attempt at reviewing it properly; the reader is advised to follow the survey articles [8, 88] and books [22, 25] (which cover way more on the theme) for this. We would nevertheless mention the original approach of J. Guillera for proving the formulae by J. Guillera [38–41] using the powerful Wilf–Zeilberger (WZ) machinery; the method in its basic form is exemplified in Exercise 10.2. Guillera manages to prove similar-looking identities for $1/\pi^2$ in terms of ${}_5F_4$ hypergeometric series, and his method (quite elementary in nature!) is currently the only one which is available for such formulae.

The method of creative microscoping originates from the paper [44]. The name 'creative microscoping' is inspired by 'creative telescoping' — the latter coined in [64] to the method which was originally used by D. Zagier for proving the recurrence equation in Apéry's proof of the irrationality of $\zeta(3)$ (see Exercise 7.2). In this chapter we have witnessed several other applications of creative telescoping.

It is worth mentioning that the congruences in (10.17) and Theorems 10.6 and 10.7 remain true when the sums are truncated at $(p-1)/2$

or $(n-1)/2$, respectively; these other(!) companion congruences can also be settled by the method. Recent work of V. Guo, some in collaboration with M. Schlosser, and with others (see, for example, [43, 45]), extends the horizons of applicability of creative microscoping even further. One of the latest achievements is a general framework (of q-analogues) of so-called Dwork-type supercongruences.

Bibliography

[1] M. AIGNER and G. ZIEGLER, *Proofs from The Book*, 6th edition (Springer, Berlin, 2018).

[2] G. E. ANDREWS and B. C. BERNDT, *Ramanujan's lost notebook. Part I* (Springer, New York, 2005); *Part II* (Springer, New York, 2009); *Part III* (Springer, New York, 2012); *Part IV* (Springer, New York, 2013); *Part V* (Springer, Cham, 2018).

[3] R. APÉRY, Irrationalité de $\zeta(2)$ et $\zeta(3)$, *Astérisque* **61** (1979), 11–13.

[4] T. M. APOSTOL, *Introduction to analytic number theory*, Undergraduate Texts in Mathematics (Springer-Verlag, New York–Heidelberg, 1976).

[5] T. ARAKAWA, T. IBUKIYAMA and M. KANEKO, *Bernoulli numbers and zeta functions*, with an appendix by D. Zagier, Springer Monographs in Math. (Springer, Tokyo, 2014).

[6] W. N. BAILEY, *Generalized hypergeometric series* (Cambridge Univ. Press, Cambridge, 1935); *Reprinted*, Cambridge Tracts in Math. and Math. Physics **32** (Stechert-Hafner, Inc., New York, 1964).

[7] A. BAKER, *Transcendental number theory*, 2nd edition, Cambridge Math. Library (Cambridge Univ. Press, Cambridge, 1990).

[8] N. D. BARUAH, B. C. BERNDT and H. H. CHAN, Ramanujan's series for $1/\pi$: a survey, *Amer. Math. Monthly* **116** (2009), 567–587.

[9] P. T. BATEMAN and H. G. DIAMOND, *Analytic number theory. An introductory course*, Monographs in Number Theory **1** (World Sci. Publ. Co. Pte. Ltd., Hackensack, NJ, 2004).

[10] B. C. BERNDT, *Ramanujan's notebooks. Part I* (Springer-Verlag, New York, 1985); *Part II* (Springer-Verlag, New York, 1989); *Part III* (Springer-Verlag, New York, 1991); *Part IV* (Springer-Verlag, New York, 1994); *Part V* (Springer-Verlag, New York, 1998).

[11] B. C. BERNDT and R. A. RANKIN, *Ramanujan. Letters and commentary*, History of Math. **9** (Amer. Math. Soc., Providence, RI & London Math. Soc., London, 1995).

[12] F. BEUKERS, A note on the irrationality of $\zeta(2)$ and $\zeta(3)$, *Bull. London Math. Soc.* **11** (1979), no. 3, 268–272.

[13] J. W. BOBER, Factorial ratios, hypergeometric series, and a family of step

functions, *J. London Math. Soc.* (2) **79** (2009), 422–444.

[14] J. M. BORWEIN, A. VAN DER POORTEN, J. SHALLIT and W. ZUDILIN, *Neverending fractions. An introduction to continued fractions*, Aust. Math. Soc. Lecture Ser. **23** (Cambridge University Press, Cambridge, 2014).

[15] F. BROWN, Irrationality proofs for zeta values, moduli spaces and dinner parties, *Mosc. J. Comb. Number Theory* **6** (2016), no. 2-3, 102–165.

[16] F. BROWN and W. ZUDILIN, On cellular rational approximations to $\zeta(5)$, *Preprint* `arXiv: 2210.03391 [math.NT]` (2022), 31 pp.

[17] N. G. DE BRUIJN, Asymptotic methods in analysis, *Bibliotheca Mathematica* **4** (North-Holland Publ. Co., Amsterdam & P. Noordhoff Ltd., Groningen, 1958).

[18] F. BRUNAULT and W. ZUDILIN, *Many variations of Mahler measures: a lasting symphony*, Aust. Math. Soc. Lecture Ser. **28** (Cambridge University Press, Cambridge, 2020).

[19] Y. BUGEAUD, *Linear forms in logarithms and applications*, IRMA Lectures in Math. and Theor. Phys. **28** (European Math. Soc., Zürich, 2018).

[20] J. I. BURGOS GIL, J. FRESÁN, with contributions by U. KÜHN, *Multiple zeta values: from numbers to motives*, Clay Mathematics Proceedings (in press); `http://javier.fresan.perso.math.cnrs.fr/mzv.pdf`.

[21] H. H. CHAN, *Analytic number theory for undergraduates*, Monographs in Number Theory **3** (World Sci. Publ. Co. Pte. Ltd., Hackensack, NJ, 2009).

[22] H. H. CHAN, *Theta functions, elliptic functions and* π, with a foreword by B. Berndt, De Gruyter Textbook (De Gruyter, Berlin, 2020).

[23] H. H. CHAN and W. ZUDILIN, New representations for Apéry-like sequences, *Mathematika* **56** (2010), no. 1, 107–117.

[24] J. S. CHRISTIANSEN, B. SIMON and M. ZINCHENKO, Asymptotics of Chebyshev polynomials. IV. Comments on the complex case, *J. Anal. Math.* **141** (2020), no. 1, 207–223.

[25] S. COOPER, *Ramanujan's theta functions* (Springer, Cham, 2017).

[26] V. DIMITROV, A proof of the Schinzel–Zassenhaus conjecture on polynomials, *Preprint* `arXiv: 1912.12545 [math.NT]` (2019), 27 pp.

[27] V. N. DUBININ, On the change in harmonic measure under symmetrization, *Mat. Sb.* **52** (1985), no. 1, 267–273.

[28] S. B. EKHAD and D. ZEILBERGER, A WZ proof of Ramanujan's formula for π, in *Geometry, Analysis, and Mechanics*, J. M. Rassias (ed.) (World Scientific, Singapore, 1994), pp. 107–108.

[29] M. FEKETE, Über die Verteilung der Wurzeln bei gewissen algebraischen Gleichungen mit ganzzahligen Koeffizlenten, *Math. Zeit.* **17** (1923), 228–249.

[30] S. FISCHLER, Irrationalité de valeurs de zêta (d'après Apéry, Rivoal, ...), *Astérisque* **294** (2004), 27–62.

[31] S. FISCHLER, J. SPRANG and W. ZUDILIN, Many odd zeta values are irrational, *Compos. Math.* **155** (2019), no. 5, 938–952.

[32] J. FRIEDLANDER and H. IWANIEC, The polynomial $X^2 + Y^4$ captures its primes, *Ann. of Math.* (2) **148** (1998), no. 3, 945–1040; Asymptotic sieve for primes, *Ann. of Math.* (2) **148** (1998), no. 3, 1041–1065.

[33] A. I. GALOCHKIN, YU. V. NESTERENKO and A. B. SHIDLOVSKII, *Introduction to number theory* [in Russian], 2nd edition (Moscow State Univ. Publ., Moscow, 1995).

[34] G. GASPER and M. RAHMAN, *Basic hypergeometric series*, 2nd edition, Encyclopedia Math. Appl. **96** (Cambridge Univ. Press, Cambridge, 2004).

[35] I. GESSEL, Super ballot numbers, *J. Symbolic Computation* **14** (1992), 179–194.

[36] A. GRANVILLE, *Number theory revealed: a masterclass* (Amer. Math. Soc., Providence, RI, 2019).

[37] B. GREEN and T. TAO, The primes contain arbitrarily long arithmetic progressions, *Ann. of Math.* (2) **167** (2008), no. 2, 481–547.

[38] J. GUILLERA, Some binomial series obtained by the WZ-method, *Adv. in Appl. Math.* **29** (2002), no. 4, 599–603.

[39] J. GUILLERA, About a new kind of Ramanujan-type series, *Experiment. Math.* **12** (2003), no. 4, 507–510.

[40] J. GUILLERA, A new Ramanujan-like series for $1/\pi^2$, *Ramanujan J.* **26** (2011), no. 3, 369–374.

[41] J. GUILLERA, WZ pairs and q-analogues of Ramanujan series for $1/\pi$, *J. Difference Equ. Appl.* **24** (2018), no. 12, 1871–1879.

[42] J. GUILLERA and W. ZUDILIN, Ramanujan-type formulae for $1/\pi$: The art of translation, in *The Legacy of Srinivasa Ramanujan*, B. C. Berndt and D. Prasad (eds.), Ramanujan Math. Soc. Lecture Notes Series **20** (2013), pp. 181–195.

[43] V. J. W. GUO and M. J. SCHLOSSER, Some new q-congruences for truncated basic hypergeometric series: even powers, *Results Math.* **75** (2020), no. 1, paper no. 1, 15 pp.

[44] V. J. W. GUO and W. ZUDILIN, A q-microscope for supercongruences, *Adv. Math.* **346** (2019), 329–358.

[45] V. J. W. GUO and W. ZUDILIN, Dwork-type supercongruences through a creative q-microscope, *J. Combin. Theory Ser. A* **178** (2021), paper no. 105362, 37 pp.

[46] P. HABEGGER, Separating roots of polynomials and the transfinite diameter, *Riv. Math. Univ. Parma (N.S.)* **13** (2022), no. 1, 111–136.

[47] H. HASSE, *Number theory*, Translated from the 3rd (1969) German edition; Reprint of the 1980 English edition, Classics in Mathematics (Springer-Verlag, Berlin, 2002).

[48] M. HATA, *Problems and solutions in real analysis*, 2nd edition, Series on Number Theory and its Applications **14** (World Sci. Publ. Co., Hackensack, NJ, 2017).

[49] D. R. HEATH-BROWN, Primes represented by $x^3 + 2y^3$, *Acta Math.* **186** (2001), no. 1, 1–84.

[50] L. KRONECKER, Zur Theorie der Elimination einer Variabeln aus zwei algebraischen Gleichungen, *Berl. Monatsber.* **1881** (1881), 535–600.

[51] L. LAI and P. YU, A note on the number of irrational odd zeta values, *Compos. Math.* **156** (2020), no. 8, 1699–1717.

[52] E. LANDAU, Sur les conditions de divisibilité d'un produit de factorielles par

un autre, in: *Collected Works*, Vol. I (Thales-Verlag, Essen, 1985), p. 116.
[53] S. Lang, Introduction to transcendental numbers (Addison-Wesley Publishing Co., Reading, Mass. & London-Don Mills, Ont., 1966).
[54] M. Laurent, Sur quelques résultats récents de transcendance, *Astérisque* **198–200** (1991), 209–230.
[55] M. Laurent, Linear forms in two logarithms and interpolation determinants, *Acta Arith.* **66** (1994), no. 2, 181–199.
[56] M. Laurent, M. Maurice and Yu. Nesterenko, Formes linéaires en deux logarithmes et déterminants d'interpolation, *J. Number Theory* **55** (1995), no. 2, 285–321.
[57] J. Maynard, Small gaps between primes, *Ann. of Math.* (2) **181** (2015), no. 1, 383–413.
[58] Yu. V. Nesterenko, Some remarks on $\zeta(3)$, *Math. Notes* **59** (1996), no. 5-6, 625–636.
[59] D. J. Newman, *Analytic number theory*, Graduate Texts in Math. **177** (Springer-Verlag, New York, 1998).
[60] K. Ono, Unearthing the visions of a master: harmonic Maass forms and number theory, in: *Current developments in mathematics*, 2008 (Intern. Press, Somerville, MA, 2009), pp. 347–454.
[61] M. Petkovšek, H. S. Wilf and D. Zeilberger, $A = B$ (A. K. Peters, Ltd., Wellesley, MA, 1997).
[62] G. Pólya, Über gewisse notwendige Determinantenkriterien für die Fortsetzbarkeit einer Potenzreihe, *Math. Ann.* **99** (1928), 687–706.
[63] G. Pólya and G. Szegő, *Problems and theorems in analysis*, Vol. II, Grundlehren Math. Wiss. **216** (Springer-Verlag, Berlin et al., 1976).
[64] A. van der Poorten, A proof that Euler missed... Apéry's proof of the irrationality of $\zeta(3)$ (An informal report), *Math. Intelligencer* **1** (1978/79), no. 4, 195–203.
[65] S. Ramanujan, Modular equations and approximations to π, *Quart. J. Math.* **45** (1914), 350–372.
[66] G. Rhin and C. Viola, The group structure for $\zeta(3)$, *Acta Arith.* **97** (2001), no. 3, 269–293.
[67] T. Rivoal, La fonction zêta de Riemann prend une infinité de valeurs irrationnelles aux entiers impairs, *C. R. Acad. Sci. Paris Ser. I* **331** (2000), no. 4, 267–270.
[68] K. F. Roth, Rational approximations to algebraic numbers, *Mathematika* **2** (1955), 1–20; corrigendum, 168.
[69] A. Schinzel and H. Zassenhaus, A refinement of two theorems of Kronecker, *Michigan Math. J.* **12** (1965), 81–85.
[70] Th. Schneider, Ein Satz über ganzwertige Funktionen als Prinzip für Transzendenzbeweise, *Math. Ann.* **121** (1949), 131–140.
[71] L. J. Slater, *Generalized hypergeometric functions* (Cambridge Univ. Press, Cambridge, 1966).
[72] C. J. Smyth, A coloring proof of a generalisation of Fermat's little theorem, *Amer. Math. Monthly* **93** (1986), no. 6, 469–471.
[73] R. P. Stanley, *Enumerative combinatorics*, vol. 2, with a foreword by

G.-C. Rota, Cambridge Studies in Advanced Mathematics **62** (Cambridge Univ. Press, Cambridge, 1999).

[74] A. STRAUB, Supercongruences for polynomial analogs of the Apéry numbers, *Proc. Amer. Math. Soc.* **147** (2019), no. 3, 1023–1036; *Extended preprint* `arXiv: 1803.07146 [math.NT]` (2018), 19 pp.

[75] R. TAYLOR, Reciprocity laws and density theorems, *Shaw lecture* (2007), 19 pp.; `http://virtualmath1.stanford.edu/~rltaylor/shaw.pdf`.

[76] P. L. TCHEBICHEF, Mémoire sur les nombres premiers, *J. Math. Pures Appl.* **17** (1852), 366–390.

[77] C. VIOLA, *An introduction to special functions*, Unitext **102** (Springer, Cham, 2016).

[78] M. WALDSCHMIDT, *Diophantine approximation on linear algebraic groups. Transcendence properties of the exponential function in several variables*, Die Grundlehren der mathematischen Wissenschaften **326** (Springer-Verlag, Berlin, 2000).

[79] S. O. WARNAAR and W. ZUDILIN, A q-rious positivity, *Aequat. Math.* **81** (2011), no. 1-2, 177–183.

[80] E. T. WHITTAKER and G. N. WATSON, *A course of modern analysis*, 4th edition (Cambridge University Press, 1927).

[81] H. S. WILF and D. ZEILBERGER, An algorithmic proof theory for hypergeometric (ordinary and "q") multisum/integral identities, *Invent. Math.* **108** (1992), no. 3, 575–633.

[82] M. YOSHIDA, *Hypergeometric functions, my love. Modular interpretations of configuration spaces*, Aspects of Math. **E32** (Friedr. Vieweg & Sohn, Braunschweig, 1997).

[83] A. V. ZARELUA, On congruences for the traces of powers of some matrices, *Proc. Steklov Inst. Math.* **263** (2008), 78–98.

[84] D. ZEILBERGER, The method of creative telescoping, *J. Symbolic Comput.* **11** (1991), no. 3, 195–204.

[85] D. ZEILBERGER and W. ZUDILIN, The irrationality measure of π is at most $7.103205334137\dots$, *Mosc. J. Comb. Number Theory* **9** (2020), no. 4, 407–419.

[86] Y. ZHANG, Bounded gaps between primes, *Ann. of Math.* (2) **179** (2014), no. 3, 1121–1174.

[87] J. ZHAO, *Multiple zeta functions, multiple polylogarithms and their special values*, Series on Number Theory and its Applications **12** (World Sci. Publ. Co. Pte. Ltd., Hackensack, NJ, 2016).

[88] W. ZUDILIN, Ramanujan-type formulae for $1/\pi$: A second wind?, in *Modular Forms and String Duality* (Banff, June 3–8, 2006), N. Yui, H. Verrill and C.F. Doran (eds.), Fields Inst. Commun. Ser. **54** (Amer. Math. Soc., Providence, RI, 2008), pp. 179–188.

[89] W. ZUDILIN, Ramanujan-type supercongruences, *J. Number Theory* **129** (2009), 1848–1857.

[90] W. ZUDILIN, One of the odd zeta values from $\zeta(5)$ to $\zeta(25)$ is irrational. By elementary means, *SIGMA Symmetry Integrability Geom. Methods Appl.* **14** (2018), paper no. 028, 8 pp.

Index

Printed in the United States
by Baker & Taylor Publisher Services